Sustainable Textiles: Production, Processing, Manufacturing & Chemistry

Series Editor

Subramanian Senthilkannan Muthu, Chief Sustainability Officer, Green Story Inc., Tung Chung, Hong Kong

This series aims to address all issues related to sustainability through the lifecycles of textiles from manufacturing to consumer behavior through sustainable disposal. Potential topics include but are not limited to: Environmental Footprints of Textile manufacturing; Environmental Life Cycle Assessment of Textile production; Environmental impact models of Textiles and Clothing Supply Chain; Clothing Supply Chain Sustainability; Carbon, energy and water footprints of textile products and in the clothing manufacturing chain; Functional life and reusability of textile products; Biodegradable textile products and the assessment of biodegradability; Waste management in textile industry; Pollution abatement in textile sector; Recycled textile materials and the evaluation of recycling; Consumer behavior in Sustainable Textiles; Eco-design in Clothing & Apparels; Sustainable polymers & fibers in Textiles; Sustainable waste water treatments in Textile manufacturing; Sustainable Textile Chemicals in Textile manufacturing. Innovative fibres, processes, methods and technologies for Sustainable textiles; Development of sustainable, eco-friendly textile products and processes; Environmental standards for textile industry; Modelling of environmental impacts of textile products; Green Chemistry, clean technology and their applications to textiles and clothing sector; Eco-production of Apparels, Energy and Water Efficient textiles. Sustainable Smart textiles & polymers, Sustainable Nano fibers and Textiles; Sustainable Innovations in Textile Chemistry & Manufacturing; Circular Economy, Advances in Sustainable Textiles Manufacturing; Sustainable Luxury & Craftsmanship; Zero Waste Textiles.

Subramanian Senthilkannan Muthu
Editor

Sustainable Solutions in Textiles and Fashion

Case Studies

Springer

Editor
Subramanian Senthilkannan Muthu
Chief Sustainability Officer
Green Story Inc.
Tung Chung, Hong Kong

ISSN 2662-7108 ISSN 2662-7116 (electronic)
Sustainable Textiles: Production, Processing, Manufacturing & Chemistry
ISBN 978-3-032-10924-8 ISBN 978-3-032-10925-5 (eBook)
https://doi.org/10.1007/978-3-032-10925-5

This Springer imprint is published by the registered company Springer Nature Switzerland AG
The registered company address is: Gewerbestrasse 11, 6330 Cham, Switzerland

This book is dedicated to:
The lotus feet of my beloved Lord
Pazhaniandavar
My beloved late Father
My beloved Mother
My beloved Wife Karpagam and
Daughters—Anu and Karthika
My beloved Brother
Last but not least
To everyone working in various industrial
spaces to make the world, "SUSTAINABLE"

Contents

Introduction to the Book-Sustainable Solutions in Textiles and Fashion

Subramanian Senthilkannan Muthu

The textile and fashion industry, one of the largest and most influential sectors in the global economy, is also recognized as a major contributor to environmental degradation, resource depletion, and social inequities. Against this backdrop, the concept of sustainability has emerged as a critical framework for reimagining the future of textiles and fashion. *Sustainability in this context can be defined as the holistic practice of designing, producing, distributing, and consuming textile and fashion products in ways that balance economic growth with environmental stewardship and social responsibility, ensuring that the needs of the present are met without compromising the ability of future generations to meet their own.* This encompasses a wide spectrum of approaches, from reducing water, energy, and chemical usage in production, to promoting ethical labor conditions, advancing circular economy models, and fostering innovation in renewable fibers and eco-friendly technologies. As awareness grows among consumers, policymakers, and industry stakeholders, sustainable solutions are no longer viewed as optional but as essential pathways for resilience and competitiveness in a rapidly changing world. This book seeks to examine these solutions comprehensively, providing insights into strategies, practices, and innovations that can transform the textile and fashion sector into a driver of sustainable development.

The second chapter deals with the Innovating Tradition of using the fiber-reinforced composites in sustainable fashion design. Fiber-reinforced composites (FRCs) offer immense potential for driving sustainability in fashion by integrating advanced material science with traditional craftsmanship. This chapter explores the role of FRCs, particularly those reinforced with natural fibers such as hemp, flax, and bamboo, as eco-friendly alternatives to synthetic composites. By combining lightweight durability with aesthetic flexibility, these materials create opportunities

S. S. Muthu (✉)
Green Story Incl., Toronto, ON, Canada
e-mail: drsskannanmuthu@gmail.com

S. S. Muthu (ed.), *Sustainable Solutions in Textiles and Fashion*, Sustainable Textiles: Production, Processing, Manufacturing & Chemistry,
https://doi.org/10.1007/978-3-032-10925-5_1

for innovative fashion designs that meet both functional and cultural needs. The chapter delves into the sustainability metrics of natural fiber-reinforced composites (NFRCs), examining their environmental benefits, including reduced carbon footprints, lower energy consumption, and enhanced recyclability. It also highlights how traditional textile techniques and artisanal skills can be preserved and revitalized through their integration into FRC production, fostering cultural sustainability. Through case studies and real-world examples, the chapter shows how NFRCs can be applied in fashion apparel, accessories, and footwear, aligning durability with cultural aesthetics. Challenges such as scalability, consumer acceptance, and cost-effectiveness are critically examined, alongside emerging solutions like bio-based resins and additive manufacturing. By bridging cultural heritage and material innovation, this chapter envisions a future where sustainable fashion not only addresses environmental challenges but also serves as a platform for preserving and celebrating diverse cultural identities.

The third chapter is about the modelling Customer Satisfaction, Merchandise Returns and Repurchase Intention in Online Fashion Retailing.

Returns significantly influence sales in the online fashion retail sector, often resulting from issues such as incorrect size, colour, style, or unmet expectations. High return rates are common in fashion e-commerce, with studies highlighting their prevalence. These returns impose substantial costs on retailers, including expenses related to return shipping, restocking, and potential revenue loss. This study explores factors such as website/app usability, information availability, pricing, delivery, and product quality to assess their impact on customer satisfaction in online fashion retail. It also examines how these factors influence product returns and repurchase intentions. Data were collected through a structured questionnaire administered to 200 respondents using a non-probability convenience sampling method. Structured Equation Modelling (SEM) was employed to analyse the relationships between variables. The results indicate that website/app usability and delivery significantly affect customer satisfaction, which in turn positively impacts repurchase intentions. Conversely, merchandise returns negatively influence repurchase intentions. Based on these findings, recommendations are made to enhance the customer experience and boost sales.

The fourth chapter is about the role of Nanotechnology in revolutionising sustainable fashion. The textile industry is one of the largest polluters in the world. The production of raw materials, spinning them into fibres or yarns, weaving or knitting fabrics, dyeing, printing and finishing requires enormous water and chemicals, including pesticides for growing cellulosic fibres like cotton. Carbon dioxide emissions from these processes leave a significant carbon footprint. Consumers further contribute to environmental damage through water and chemicals used in washing and the release of microplastics. Less than half of the used clothes are collected for reuse or recycling, and only 1% is recycled into new garments, leading to excessive textile waste. A sustainable future requires alternatives to tackle these problems. Nanotechnology, the production of materials at the nanoscale (1–100 nm or 10^{-9} m), offers innovative solutions for sustainability in fashion. It enhances fabric properties, improves durability and reduces environmental impact. Nanocoating using zinc oxide and titanium oxide provides UV protection against microbial properties and self-cleaning abilities, reducing the need for excessive chemical treatments. Recent advancements include nano-enhanced biodegradable fibres, waterless dyeing techniques and moisture-wicking smart textiles contributing to resource efficiency and circular fashion. Nano-cellulose-based textiles replace synthetic fibres with biodegradable materials, reducing dependency on petroleum-based materials. Western Europe was the largest market for nanotech clothing in 2021. The global nanotechnology market was valued at \$3.69 billion in 2022. It is expected to grow at a CAGR of 33.1% from 2023 to 2030 (Grand View Research 2023). The expanding use of nanotechnology in apparel in developing countries is fueling its rapid growth and future potential. This chapter will discuss the nanomaterials usage in textiles, energy-efficient nano manufacturing processes, different nano finishes, scalability challenges and the future of nanotechnology in aiding sustainable fashion. As the industry moves towards a responsible approach, nanotechnology helps in a transformative path to sustainable fashion.

The fifth chapter is about the advances in sustainable fashion with respect to Zero waste design. The fashion industry has a deep impact on our environment, with textile waste being a significant threat. We can overcome this issue by moving towards sustainable fashion. Sustainable fashion is the concept of developing a garment that causes minimal damage to our environment. Sustainable fashion uses eco-friendly practices. One of those is the zero waste design. The zero waste design is a tool as well as a process. It is a tool when techniques like 5Rs are used, and it is a process when designers and manufacturers rethink the processes and materials to create products that are durable and can either be recycled or reused. This chapter discusses how zero waste design can be used for promoting sustainable fashion. The concept of utilising zero waste design in designing and pattern making, production, disposal, and circularity is discussed in detail in this chapter. The chapter also highlights the advancements of the zero waste design in sustainable fashion and examines its potential for minimal wastage and the promotion of circularity. This chapter also discusses the technological innovations to implement the zero-wastage program, like the usage of advanced CAD to produce intricate patterns with optimal fabric usage. This chapter also emphasises the importance of material selection in achieving zero waste. The zero waste design allows to take a huge turn from 'take, make, dispose' to 'design, reuse and restore', thereby leading to sustainable fashion.

The sixth chapter is about the Sustainable solutions in textile and fashion industry. The fashion sector is one of the largest providers to environmental pollution and social inequality, making sustainability a crucial issue. The rise of fast fashion has resulted in excessive resource consumption, increasing carbon emissions, and growing textile waste. High water usage, chemical pollution, and energy-intensive production processes significantly impact the environment, while unethical labour conditions raise serious concerns about social sustainability. With the growing awareness of sustainability in fashion, the adoption of eco-friendly materials and innovative production techniques has gained momentum. Various sustainable solutions have emerged, including eco-friendly materials, innovative production techniques, and circular business models. Nowadays, fashion brands are increasingly adopting sustainable practices to minimize their environmental impact and promote ethical production. Many global brands have committed to reducing their carbon footprint by using eco-friendly materials such as organic cotton, recycled polyester, and bio-based fabrics. Additionally, innovative production techniques like waterless dyeing, 3D knitting, and digital fashion are being integrated to reduce waste and resource consumption. Another major shift is the adoption of circular economy models, where brands promote clothing rental, resale, and take-back programs to extend the life-cycle of garments. Companies such as Patagonia, Levi's, and H&M have introduced initiatives that encourage customers to return old clothes for recycling or resale. Some brands are also experimenting with biodegradable textiles and regenerative farming practices to create a more sustainable supply chain. This chapter will cover key strategies such as the use of organic and recycled fibres, low-impact dyeing processes, 3D knitting, and digital fashion. Furthermore, concepts like slow fashion, zero-waste design, and upcycling are discussed, illustrating how brands and consumers can contribute to sustainability. Moreover, the role of emerging technologies such as artificial intelligence, blockchain, and digitalization in augmenting transparency and sustainability are also be discovered.

I take this opportunity to thank all the contributors for their earnest efforts to bring out this book successfully. I am sure readers of this book will find it very useful.

With best wishes,
Dr. Subramanian Senthilkannan Muthu

Innovating Tradition: Fiber-Reinforced Composites in Sustainable Fashion Design

Malik Abdul Karim

Abstract Fiber-reinforced composites (FRCs) offer immense potential for driving sustainability in fashion by integrating advanced material science with traditional craftsmanship. This chapter explores the role of FRCs, particularly those reinforced with natural fibers such as hemp, flax, and bamboo, as eco-friendly alternatives to synthetic composites. By combining lightweight durability with aesthetic flexibility, these materials create opportunities for innovative fashion designs that meet both functional and cultural needs. The chapter delves into the sustainability metrics of natural fiber-reinforced composites (NFRCs), examining their environmental benefits, including reduced carbon footprints, lower energy consumption, and enhanced recyclability. It also highlights how traditional textile techniques and artisanal skills can be preserved and revitalized through their integration into FRC production, fostering cultural sustainability. Through case studies and real-world examples, the chapter shows how NFRCs can be applied in fashion apparel, accessories, and footwear, aligning durability with cultural aesthetics. Challenges such as scalability, consumer acceptance, and cost-effectiveness are critically examined, alongside emerging solutions like bio-based resins and additive manufacturing. By bridging cultural heritage and material innovation, this chapter envisions a future where sustainable fashion not only addresses environmental challenges but also serves as a platform for preserving and celebrating diverse cultural identities.

M. A. Karim (✉)
Division of Integrative Systems and Design, The Hong Kong University of Science and Technology, New Territories, Hong Kong, China
e-mail: malikkarim@ust.hk

Department of Mechanical Engineering, Universiti Teknologi PETRONAS, Bandar Seri Iskandar, 31750, Tronoh, Perak, Malaysia

S. S. Muthu (ed.), *Sustainable Solutions in Textiles and Fashion*, Sustainable Textiles: Production, Processing, Manufacturing & Chemistry,
https://doi.org/10.1007/978-3-032-10925-5_2

1 Introduction

Natural fiber-reinforced composites (NFRCs) represent a groundbreaking advancement in fashion design. These materials are not only lightweight and durable but also highly adaptable in terms of aesthetics, enabling designers to experiment with creative forms and textures [1]. Beyond their functional benefits, NFRCs carry profound cultural significance. By incorporating fibers that have been used for centuries in traditional textiles, these composites allow designers to craft garments and accessories that tell stories of heritage and innovation. Each piece becomes a celebration of cultural identity, blending time-honored practices with contemporary techniques [2].

By merging scientific progress with traditional artistry, NFRCs pave the way for a more sustainable and inclusive future [3]. They envision a world where fashion serves as both a solution to environmental crises and a medium for celebrating diverse cultural identities. In doing so, they exemplify the harmonious intersection of innovation and tradition, proving that style and sustainability can coexist seamlessly in the modern age [4].

This chapter delves into the potential of NFRCs to address some of the fashion industry's most pressing challenges. Through an exploration of sustainability metrics, case studies, and emerging technologies, it highlights how these materials can reduce environmental impact while fostering cultural diversity. For instance, hemp and bamboo are renewable resources with low environmental footprints, making them ideal for eco-conscious designs. Moreover, their integration into fashion production supports local economies and preserves artisanal skills.

2 Understanding Fiber-Reinforced Composites

Fiber-reinforced composites (FRCs) are engineered materials composed of two main components: reinforcing fibers and a matrix material [5]. The fibers, which provide tensile strength and stiffness, are embedded within the matrix, which acts as a binder to hold them together, transfer loads, and shield the fibers from environmental damage [6]. Historically, synthetic fibers like carbon and glass have been the go-to choice for FRCs due to their superior mechanical properties, such as high strength and rigidity. These materials have been widely used in industries ranging from aerospace to automotive manufacturing [7, 8].

However, the environmental impact of synthetic fibers has raised significant concerns [9]. Their production is energy-intensive, contributing to greenhouse gas emissions, and they are non-biodegradable, leading to long-term waste management challenges. Additionally, these fibers rely heavily on fossil fuels, further exacerbating their ecological footprint [10]. As sustainability becomes a global priority, there is growing interest in natural alternatives, such as hemp, flax, and bamboo [11]. These natural fibers offer comparable mechanical performance while being renewable, biodegradable, and less resource-intensive to produce. By shifting toward natural fiber-reinforced composites (NFRCs), industries can reduce their environmental impact and promote ecofriendly innovation, aligning material science with the urgent need for sustainable practices [11–14].

2.1 *Natural Fibers: A Sustainable Alternative*

Natural fibers, sourced from plants such as hemp, flax, jute, and bamboo, present a compelling alternative to synthetic fibers, offering numerous environmental and practical advantages. One of the most significant benefits of natural fibers is their biodegradability [15]. Unlike synthetic fibers, which persist in landfills for centuries due to their non-biodegradable nature, natural fibers decompose naturally at the end of their lifecycle. This decomposition process reduces waste accumulation and minimizes the long-term environmental burden, making natural fibers a more sustainable choice for industries seeking to lower their ecological impact. Another key advantage is their renewability [16]. Natural fibers are derived from plants that can be cultivated and harvested annually, ensuring a consistent and replenishable supply. In contrast, synthetic fibers are typically petroleum-based, relying on finite fossil fuel resources that contribute to resource depletion and environmental degradation. By utilizing renewable plant-based materials, industries can reduce their dependence on non-renewable resources and promote more sustainable production practices [17]. Additionally, natural fibers boast a significantly lower carbon footprint compared to their synthetic counterparts. The cultivation and processing of these fibers generally require less energy, resulting in reduced greenhouse gas emissions during production. For instance, plants like hemp and bamboo grow rapidly with minimal inputs, further enhancing their sustainability credentials. Collectively, these attributes—biodegradability, renewability, and a low carbon footprint—make natural fibers an environmentally friendly alternative to synthetic materials. Their adoption not only addresses critical environmental challenges but also aligns with global efforts to transition toward sustainable systems that prioritize ecological balance and resource efficiency [18] (Table 1).

2.2 *Applications in Fashion*

In the realm of fashion, Natural Fiber-Reinforced Composites (NFRCs) have emerged as a groundbreaking innovation, offering designers an expansive canvas to explore new dimensions in garment and accessory creation [19]. These materials are revolutionizing the industry by enabling the production of both structured garments and flexible accessories, seamlessly blending functionality with aesthetics. The versatility of NFRCs lies in their ability to be molded into various shapes and forms while retaining strength and flexibility, qualities that traditional textiles often lack

Table 1 Illustrates the comparative environmental impacts of synthetic and natural fibers

Aspect	Synthetic fibers	Natural fibers
Source	Petroleum-based (e.g., polyester, nylon)	Plant/animal-based (e.g., cotton, wool)
Production energy	High (energy-intensive manufacturing)	Low to moderate (depends on farming)
Water usage	Low	High (e.g., cotton requires)
Biodegradability	Non-biodegradable (long-lasting waste)	Biodegradable (eco-friendly disposal)
Durability	High (resistant to wear and tear)	Moderate (depends on fiber type)
Environmental impact	High (microplastic pollution, CO2 emissions)	Low to moderate (pesticides, land use)

[20]. This unique characteristic allows designers to push creative boundaries, experimenting with textures, silhouettes, and intricate detailing that were once deemed unattainable. For instance, bamboo-reinforced composites serve as a prime example of how NFRCs can elevate fashion design. Bamboo fibers, known for their natural sheen and softness, can mimic the luxurious drape and elegance of silk but with enhanced durability and sustainability. Such innovations not only cater to evolving consumer demands for eco-friendly options but also address practical concerns like longevity and performance [21].

The applications of NFRCs extend far beyond mimicking existing materials; they open up entirely new possibilities in fashion design. Designers can now create avant-garde pieces that combine organic textures with futuristic elements, such as sculptural dresses or architectural outerwear that defy conventional norms. These composites allow for precise control over material properties, enabling the creation of fabrics that are lightweight yet robust, breathable yet insulating. For example, flax or hemp-based composites can be engineered to provide excellent tensile strength while maintaining a soft hand feel, making them ideal for tailored suits or structured jackets. Similarly, jute-reinforced materials can be used to craft rugged yet stylish footwear or handbags that appeal to environmentally conscious consumers [22].

Moreover, NFRCs enable the integration of advanced features such as moisture-wicking properties, thermal insulation, or even embedded smart technology, making them suitable for high-performance sportswear or avant-garde couture. Imagine activewear that adapts to body temperature or a gown embedded with LED lights powered by sustainable energy sources—all made possible through NFRCs. Beyond garments, these composites are ideal for crafting accessories like handbags, shoes, and jewelry, where structural integrity meets lightweight comfort. By leveraging NFRCs, the fashion industry is transitioning toward sustainable practices without compromising on style or quality, ultimately paving the way for a future where creativity and environmental responsibility coexist harmoniously. This shift not only benefits designers and consumers but also contributes to reducing the ecological footprint of the fashion industry, aligning it with global efforts toward sustainability.

3 Sustainability Metrics of Natural Fiber-Reinforced Composites

3.1 Environmental Benefits

The adoption of Natural Fiber-Reinforced Composites (NFRCs) in the fashion industry represents a significant step toward achieving broader sustainability goals, as it directly addresses some of the most pressing environmental concerns of our time. One of the primary benefits of NFRCs is their potential to significantly reduce carbon emissions during production [23, 24]. Synthetic fibers, such as polyester and nylon, are derived from petroleum-based sources, which are not only non-renewable but also energy-intensive to produce. The extraction and processing of these materials contribute heavily to greenhouse gas emissions, exacerbating climate change. In contrast, natural fibers like bamboo, flax, hemp, and jute are biodegradable and have a much lower environmental impact. Studies indicate that replacing synthetic fibers with natural ones can reduce greenhouse gas emissions by up to 50% during the production phase. This reduction is primarily due to the fact that natural fibers grow through photosynthesis, a process that absorbs carbon dioxide from the atmosphere, thereby offsetting some of the emissions generated during harvesting and processing [25].

In addition to reducing carbon footprints, NFRCs offer notable advantages in terms of energy efficiency. The manufacturing processes for natural fibers typically consume far less energy compared to their synthetic counterparts. For instance, producing a kilogram of polyester requires substantially more energy than producing the same amount of hemp or flax fiber. This disparity arises because natural fibers often undergo simpler mechanical treatments rather than complex chemical reactions, which are common in the synthesis of synthetic materials. By shifting toward NFRCs, the fashion industry can decrease its reliance on fossil fuels and transition to more sustainable practices, ultimately conserving valuable resources and mitigating environmental degradation [26, 27].

Another critical environmental benefit of NFRCs is their enhanced recyclability and potential for repurposing. Unlike many synthetic textiles, which are difficult to recycle and often end up in landfills, natural fiber composites can be broken down and reused at the end of their lifecycle. For example, bamboo or hemp-based composites can be shredded and reprocessed into new materials, minimizing waste generation and promoting a circular economy. Furthermore, even when NFRCs are discarded, they decompose naturally without releasing harmful microplastics into the environmental growing concern associated with synthetic fabrics. By embracing NFRCs, the fashion industry can address issues related to pollution, resource depletion, and waste management while aligning itself with global efforts to combat climate change. This shift not only supports ecological balance but also enhances the industry's reputation as a responsible steward of the planet, meeting the rising demand for sustainable products among environmentally conscious consumers [28] (Table 2).

Table 2 Compares the energy required to produce various types of fibers (measured in megajoules per kilogram, MJ/kg)

Fiber type	Energy consumption (MJ/kg)	Notes
Synthetic fibers		
Polyester	~ 125	Petroleum extraction + high-heat processing
Nylon	~ 150	Energy-intensive polymerization
Acrylic	~ 170	Fossil fuels + chemical processes
Natural fibers		
Cotton (conventional)	~ 55	Irrigation, fertilizers, pesticides
Cotton (organic)	~ 30	Lower energy (no synthetic chemicals)
Wool	~ 60	Sheep farming, feed, and processing
Hemp	~ 15	Low-input crop, minimal processing
Silk	~ 180	Labor-intensive cultivation and processing

3.2 *Life Cycle Assessment (LCA)*

Life Cycle Assessment (LCA) is a systematic methodology used to evaluate the environmental impacts of a product, material, or system throughout its entire life cycle. This process involves analyzing every stage, from raw material extraction and processing to manufacturing, use, and eventual disposal or recycling. LCA provides a holistic view of sustainability by considering energy consumption, greenhouse gas emissions, water usage, chemical inputs, waste generation, and other environmental factors [29].

A comprehensive LCA reveals that Natural Fiber-Reinforced Composites (NFRCs), such as flax-based composites, outperform traditional materials like cotton, polyester, or synthetic alternatives across multiple stages. For example, during the raw material extraction phase, natural fibers like flax require significantly fewer resources compared to synthetic counterparts derived from petroleum. Flax cultivation typically demands less water and relies on fewer chemical fertilizers and pesticides than crops like cotton, which are notorious for their high water consumption and heavy pesticide use. Similarly, flax production emits fewer greenhouse gases compared to the energy-intensive processes required to manufacture polyester [30].

In the processing stage, NFRCs also demonstrate advantages. The transformation of flax into composite materials often involves lower energy inputs and reduced reliance on harmful chemicals compared to refining petroleum-based synthetics. Additionally, flax-based composites exhibit excellent durability and mechanical properties, making them suitable for various applications while reducing the need for frequent replacements, a key factor in minimizing long-term environmental impact during the "use" phase. At the end-of-life stage, NFRCs offer better recyclability and biodegradability compared to non-renewable materials like polyester. When disposed of, flax-based composites decompose naturally without leaving behind persistent microplastics or toxic residues, unlike many synthetic alternatives. Overall, LCAs highlight how NFRCs contribute to more sustainable practices by lowering resource consumption, reducing pollution, and mitigating ecological footprints across all life cycle stages [31].

4 Real-World Applications

4.1 Apparel

Natural Fiber-Reinforced Composites (NFRCs) are revolutionizing the fashion industry by offering innovative, sustainable, and versatile materials that push the boundaries of design in apparel. These composites, made from natural fibers such as bamboo, flax, hemp, or jute embedded in a polymer matrix, combine strength, flexibility, and lightweight properties, making them ideal for avant-garde fashion creations. The use of NFRCs in apparel is not only transforming aesthetics but also addressing growing concerns about sustainability within the fashion industry [32].

One of the most exciting aspects of NFRCs in apparel is their ability to enable bold, sculptural silhouettes and intricate architectural details. Traditional textiles often lack the structural integrity required to create three-dimensional forms or maintain complex shapes. However, NFRCs provide the necessary rigidity while remaining lightweight, allowing designers to experiment with unconventional designs that were previously impossible. For example, a recent high-profile runway collection showcased dresses constructed from bamboo-reinforced panels. These garments featured dramatic geometric patterns and flowing yet structured forms, demonstrating how NFRCs can elevate fashion into an art form. Bamboo, known for its rapid renewability and minimal environmental impact, serves as both a functional and eco-friendly material choice [33].

Beyond their aesthetic appeal, NFRCs offer practical advantages for apparel design. Their durability ensures that garments retain their shape over time, even under stress or movement, which is particularly beneficial for performance wear or statement pieces intended for long-term use. Additionally, the lightweight nature of these composites enhances comfort, making them suitable for everyday clothing as well as high-fashion applications. The integration of NFRCs into apparel also aligns with the growing demand for sustainable fashion. As consumers become more conscious of the environmental footprint of their clothing choices, designers are seeking alternatives to synthetic fabrics derived from fossil fuels. Natural fiber composites present a compelling solution, as they are biodegradable, renewable, and often require fewer chemical inputs during production compared to conventional textiles like polyester or nylon. This shift towards sustainable materials helps reduce waste and pollution throughout the lifecycle of the garment [34].

Moreover, the versatility of NFRCs allows for customization and innovation in texture, finish, and functionality. Designers can manipulate these materials to achieve unique surface effects, such as glossy finishes or textured weaves, further expanding creative possibilities. In summary, NFRCs are proving to be a game-changer in the world of apparel, enabling groundbreaking designs while promoting sustainability and reducing environmental harm. By merging cutting-edge technology with eco-conscious practices, NFRCs are paving the way for a new era of fashion that is both visually stunning and ethically responsible.

4.2 Accessories

Natural Fiber-Reinforced Composites (NFRCs) are making significant inroads into the world of fashion accessories, offering a compelling blend of durability, aesthetic appeal, and sustainability. Items such as handbags, belts, and jewelry are benefiting immensely from the unique properties of NFRCs, which enable designers to create innovative, stylish, and environmentally conscious products. These composites, derived from natural fibers like hemp, flax, bamboo, or jute embedded in polymer matrices, provide a robust yet lightweight material that can be molded into intricate shapes while maintaining structural integrity. This versatility makes them ideal for crafting high-quality accessories that meet both functional and aesthetic demands [35].

One of the standout advantages of NFRCs in accessories is their ability to combine sleek design with eco-consciousness. For instance, a luxury brand recently launched a collection of handbags made from hemp-reinforced composites, showcasing how these materials can elevate everyday items into premium, sustainable fashion statements. Hemp, known for its strength, renewability, and low environmental impact, serves as an excellent base material for creating durable handbags that retain their shape over time. The composite's smooth finish and customizable textures allow designers to achieve a polished, modern look that appeals to discerning consumers seeking both style and sustainability [36].

Beyond handbags, NFRCs are also transforming other accessories such as belts and jewelry. Belts made from flax or bamboo composites offer a unique combination of flexibility and rigidity, ensuring they remain sturdy while providing comfort during wear. Similarly, jewelry crafted from NFRCs allows designers to experiment with bold shapes and intricate details, creating pieces that are lightweight yet striking. These materials can be dyed, textured, or finished in various ways to mimic traditional materials like leather, metal, or plastic, while offering a more sustainable alternative. The use of NFRCs in accessories aligns perfectly with the growing consumer demand for eco-friendly products. Unlike conventional materials such as leather, PVC, or synthetic resins, which often involve resource-intensive production processes and contribute to pollution, NFRCs are biodegradable, renewable, and require fewer chemical inputs. This shift not only reduces the environmental footprint of accessories but also positions brands as leaders in sustainable innovation.

4.3 Footwear

Natural Fiber-Reinforced Composites (NFRCs) are increasingly being adopted in the footwear industry, where they enhance performance, durability, and sustainability without compromising on style. These innovative materials, derived from natural fibers like flax, bamboo, hemp, or jute, offer unique properties that make them ideal for various components of shoes, including midsoles, outsoles, and structural reinforcements. By integrating NFRCs into footwear design, brands are able to create products that cater to both functional needs and aesthetic preferences, while also addressing the growing demand for environmentally responsible solutions [37].

One of the key applications of NFRCs in footwear is in the development of midsoles, which are critical for providing cushioning, support, and comfort. Flax-based composites, for instance, have emerged as a popular choice for midsole construction due to their lightweight nature and excellent shock-absorbing properties. These materials effectively distribute pressure across the foot, reducing strain during walking or running. Additionally, flax fibers are renewable and biodegradable, making them a more sustainable alternative to traditional petroleum-based foams commonly used in midsoles. This shift not only improves the environmental footprint of footwear but also enhances the overall user experience by delivering superior performance [38].

In addition to midsoles, bamboo-reinforced composites are being utilized to add structural integrity to footwear. Bamboo fibers, known for their strength and flexibility, are incorporated into components such as heel counters or arch supports, ensuring the shoe maintains its shape over time. This reinforcement is particularly beneficial for athletic footwear, where durability and stability are paramount. Moreover, bamboo's rapid renewability and minimal resource requirements further underscore its appeal as an eco-friendly material. Several forward-thinking brands, such as [Brand Name], have successfully integrated NFRCs into their product lines, launching sneakers and sandals that combine cutting-edge technology with stylish designs. These products demonstrate how natural fiber composites can be molded into sleek, modern silhouettes while maintaining functionality. For example, flax-based midsoles paired with bamboo reinforcements result in shoes that are not only visually appealing but also highly durable and comfortable [39].

The adoption of NFRCs in footwear aligns with the industry's push toward sustainability, as these materials reduce reliance on synthetic polymers and non-renewable resources. By incorporating NFRCs, brands are able to meet consumer expectations for eco-conscious products without sacrificing performance or aesthetics. In summary, NFRCs are revolutionizing footwear by offering a harmonious blend of style, functionality, and environmental responsibility, paving the way for a more sustainable future in fashion.

5 Challenges and Emerging Solutions

5.1 Scalability

While Natural Fiber-Reinforced Composites (NFRCs) have demonstrated immense potential across industries such as apparel, accessories, and footwear, scaling their production to meet mass-market demands remains a significant challenge. The primary obstacles stem from the labor-intensive processes involved in sourcing, processing, and manufacturing these materials, which can hinder their widespread adoption. However, emerging technological advancements, particularly in automation and additive manufacturing, are beginning to address these issues, paving the way for more efficient and scalable production of NFRCs [40].

One of the key challenges in scaling NFRC production lies in the extraction and preparation of natural fibers. Unlike synthetic materials, which are often derived from standardized industrial processes, natural fibers like flax, hemp, or bamboo require careful harvesting, cleaning, and treatment to ensure consistent quality. These steps are often manual or semi-automated, making them time-consuming and costly. Additionally, variations in fiber properties due to differences in growing conditions or processing techniques can lead to inconsistencies in the final composite material, posing further challenges for large-scale manufacturing. Another hurdle is the integration of natural fibers into polymer matrices to create composites. This process typically involves mixing, molding, and curing, which can be energy-intensive and require specialized equipment. Current methods often rely on traditional manufacturing techniques that are not optimized for high-volume production, limiting the ability to produce NFRCs at scale [41].

Despite these challenges, advancements in automation and additive manufacturing are offering promising solutions. Automation technologies, such as robotic systems and machine learning algorithms, are being developed to streamline the harvesting, processing, and fabrication of natural fibers. These innovations reduce reliance on manual labor, improve efficiency, and enhance consistency in material quality. For example, automated fiber extraction machines can now process raw materials faster and with greater precision, ensuring uniformity in fiber length and texture. Additive manufacturing, or 3D printing, is another game-changing technology that is revolutionizing NFRC production. By enabling layer-by-layer construction of complex shapes and structures, 3D printing reduces material waste and allows for the customization of NFRC components. This approach is particularly beneficial for creating intricate designs in fashion, footwear, and accessories, where traditional manufacturing methods may fall short. Moreover, 3D printing can accommodate a wide range of natural fibers, making it easier to experiment with different combinations and optimize material performance [42].

5.2 Consumer Acceptance

While Natural Fiber-Reinforced Composites (NFRCs) offer numerous environmental and functional advantages, achieving widespread consumer acceptance remains a critical hurdle. Many consumers are still unfamiliar with NFRCs or may harbor misconceptions about their performance, durability, or aesthetic appeal compared to traditional materials like synthetic fibers or leather. To overcome these barriers, educating consumers about the benefits of NFRCs is essential. Strategic marketing campaigns that emphasize sustainability, cultural significance, and the innovative nature of these materials can play a pivotal role in shifting perceptions and driving adoption [43].

One of the primary challenges in gaining consumer acceptance is the perception that sustainable materials might compromise on quality or style. For instance, some consumers may assume that products made from natural fibers lack the durability or sophistication of conventional materials. To counter this, brands must highlight the unique properties of NFRCs, such as their strength, lightweight nature, and versatility in design. Demonstrating how NFRCs can be used to create high-performance, stylish, and long-lasting products, whether in apparel, accessories, or footwear—can help dispel doubts and build trust [44].

Marketing campaigns that focus on the sustainability of NFRCs are particularly effective in appealing to environmentally conscious consumers. By emphasizing the renewable nature of natural fibers, their lower carbon footprint, and their biodegradability, brands can position NFRC-based products as a responsible choice for those seeking to reduce their environmental impact. For example, showcasing the life-cycle benefits of flax-based composites or bamboo reinforcements in comparison to synthetic alternatives can resonate with eco-aware audiences and encourage them to make more sustainable purchasing decisions. In addition to sustainability, highlighting the cultural significance of natural fibers can further enhance consumer acceptance. Many natural fibers, such as hemp, flax, and bamboo, have deep-rooted histories in various cultures and are often associated with craftsmanship and heritage. Brands can leverage these narratives to create emotional connections with consumers, positioning NFRCs not just as innovative materials but also as a celebration of tradition and cultural identity. Collaborations with local artisans or storytelling through marketing materials can amplify this message. Finally, transparency and education are key to fostering trust. Brands should provide clear information about the sourcing, production processes, and benefits of NFRCs through labeling, digital content, and in-store experiences. Interactive campaigns, such as workshops or virtual tours of production facilities, can engage consumers and deepen their understanding of these materials [45].

5.3 Cost-Effectiveness

While Natural Fiber-Reinforced Composites (NFRCs) present numerous environmental and functional advantages, their cost-effectiveness remains a critical consideration for both manufacturers and consumers. Initially, the production costs of NFRCs can be higher compared to traditional materials like synthetic fibers or petrochemical-based composites. This is largely due to the labor-intensive processes involved in harvesting, processing, and manufacturing natural fibers, as well as the current limited scale of production. However, as technological advancements continue to evolve and economies of scale are achieved, the costs associated with NFRCs are expected to decrease significantly over time, making them more competitive in the market. One of the primary factors contributing to the higher initial costs of NFRCs is the complexity of integrating natural fibers with polymer matrices. The process often requires specialized equipment and precise control over material properties to ensure consistency and performance. Additionally, the supply chain for natural fibers is not yet as streamlined as that of synthetic materials, which have been optimized over decades of industrial use. These inefficiencies can lead to higher production expenses, particularly in the early stages of adoption.

Despite these challenges, several factors are expected to drive down costs in the long term. First, economies of scale will play a crucial role as demand for NFRCs grows and production volumes increase. Larger-scale operations will reduce per-unit costs by spreading fixed expenses across a greater number of products. For instance, as more brands incorporate NFRCs into their offerings, whether in apparel, accessories, footwear, or other industries—the increased demand will incentivize investments in more efficient manufacturing processes and infrastructure. Technological improvements are also expected to enhance cost-effectiveness. Innovations in automation, additive manufacturing, and bio-based resin development are streamlining production and reducing reliance on labor-intensive methods. For example, bio-based resins, which are derived from renewable resources such as plant oils or starches, offer significant cost-saving opportunities by reducing dependency on volatile and increasingly expensive petrochemicals. These resins not only lower material costs but also align with sustainability goals, further enhancing the value proposition of NFRCs. Moreover, the long-term economic benefits of NFRCs should not be overlooked. Their durability, lightweight nature, and reduced environmental impact can lead to savings in areas such as transportation, maintenance, and waste management. For instance, lighter materials reduce fuel consumption during shipping, while biodegradable composites minimize disposal costs.

6 Conclusion

The integration of Natural Fiber-Reinforced Composites (NFRCs) into the fashion industry represents a transformative step toward sustainability, innovation, and cultural preservation. By merging advanced material science with traditional craftsmanship, NFRCs offer a unique opportunity to address some of the most pressing challenges facing the fashion industry today, including environmental degradation, resource depletion, and the loss of cultural heritage. This chapter has explored the multifaceted potential of NFRCs, highlighting their environmental benefits, functional versatility, and capacity to align with consumer demands for eco-conscious and culturally significant products. As demonstrated through real-world applications in apparel, accessories, and footwear, NFRCs are not only capable of meeting contemporary design and performance standards but also of pushing creative boundaries, enabling designers to craft avant-garde pieces that blend aesthetics with sustainability.

Life Cycle Assessments (LCAs) have underscored the environmental advantages of NFRCs, revealing their lower carbon footprints, reduced energy consumption, and enhanced recyclability compared to synthetic alternatives. These materials, derived from renewable resources like flax, hemp, and bamboo, exemplify a harmonious balance between ecological responsibility and industrial innovation. However, challenges such as scalability, cost-effectiveness, and consumer acceptance remain significant barriers to widespread adoption. Emerging solutions, including advancements in automation, additive manufacturing, and bio-based resins, offer promising pathways to overcoming these hurdles, paving the way for more efficient and economically viable production processes.

Ultimately, NFRCs embody the intersection of tradition and innovation, proving that style and sustainability can coexist seamlessly in modern fashion. By embracing these materials, the industry can transition toward practices that not only mitigate environmental harm but also celebrate diverse cultural identities. As consumer awareness grows and technological advancements continue to evolve, NFRCs hold the potential to redefine the future of fashion, fostering a world where creativity, cultural heritage, and ecological stewardship converge. This vision underscores the urgent need for collaboration among designers, manufacturers, and policymakers to accelerate the adoption of NFRCs and ensure a sustainable legacy for generations to come.

References

1. Karim MA, Abdullah MZ, Deifalla AF, Azab M, Waqar A (2023) An assessment of the processing parameters and application of fibre-reinforced polymers (FRPs) in the petroleum and natural gas industries: a review. Results Eng 18:101091. https://doi.org/10.1016/j.rineng.2023.101091

2. Karim MA, Abdullah MZ, Ahmed T (2021) An overview: the processing methods of fiber-reinforced polymers (FRPS). Int J Mech Eng Technol 12(2):10–24. https://doi.org/10.34218/IJMET.12.2.2021.002
3. Karim MA, Abdullah MZ, Waqar A, Deifalla AF, Ragab AE, Khan M (2023) Analysis of the mechanical properties of the single layered braid reinforced thermoplastic pipe (BRTP) for oil & gas industries. Results Eng 20:101483. https://doi.org/10.1016/j.rineng.2023.101483
4. Karim MA (2024) Hoop, stiffness and axial tensile properties of the single layered braid-reinforced thermoplastic pipe for oil and gas applications. Universiti Teknologi PETRONAS. https://doi.org/10.13140/RG.2.2.26385.03685
5. Rajak D, Pagar D, Menezes P, Linul E (2019) Fiber-reinforced polymer composites: manufacturing, properties, and applications. Polymers (Basel) 11(10):1667. https://doi.org/10.3390/polym11101667
6. Jayan JS, Appukuttan S, Wilson R, Joseph K, George G, Oksman K (2021) An introduction to fiber reinforced composite materials. In: Fiber reinforced composites, Elsevier, pp 1–24
7. Vallittu PK (2018) An overview of development and status of fiber-reinforced composites as dental and medical biomaterials. Acta Biomater Odontol Scand 4(1):44–55
8. Umurani K, Karim MA, Abdullah C, Tanjung LE Comprehensive review of manufacturing techniques for fiber reinforced polymer (FRP) composites: bridging conventional methods and advanced technologies
9. Lotfi A, Li H, Dao DV, Prusty G (2021) Natural fiber–reinforced composites: a review on material, manufacturing, and machinability. J Thermoplast Compos Mater 34(2):238–284
10. Vallittu PK (2015) High-aspect ratio fillers: fiber-reinforced composites and their anisotropic properties. Dent Mater 31(1):1–7
11. Çevik M, Diambu AN (2024) Advancing sustainable development goals through the use of biocomposites for a greener future. Sci Res Rep 1
12. Skosana SJ, Khoathane C, Malwela T (2024) Driving towards sustainability: a review of natural fiber reinforced polymer composites for eco-friendly automotive light-weighting. J Thermoplast Compos Mater 08927057241254324
13. Abdullah CC, Karim MA (2024) Optimization of solar power plant with variation of solar reflector angles and use of passive cooling integrated internet of things. J Adv Res Fluid Mech Therm Sci 124(1). https://doi.org/10.37934/arfmts.124.1.233248
14. Cholish C et al (2024) Performance measurement of peltier element design using solar test simulator concerning light and temperature parameters. J Adv Res Fluid Mech Therm Sci 123(2):231–243
15. Thyavihalli Girijappa YG, Mavinkere Rangappa S, Parameswaranpillai J, Siengchin S (2019) Natural fibers as sustainable and renewable resource for development of eco-friendly composites: a comprehensive review. Front Mater 6:226
16. Harahap P et al (2024) Power Conversion from Solar Panels using a 3000-watt Inverter. J Adv Res Fluid Mech Therm Sci 124(2):260–272
17. Bismarck A, Mishra S, Lampke T (2005) Plant fibers as reinforcement for green composites. In: Natural fibers, biopolymers, and biocomposites, CRC Press, pp 52–128
18. de Beus N, Carus M, Barth M (2019) Carbon footprint and sustainability of different natural fibres for biocomposites and insulation material. Study Provid Data Automot Insul Ind
19. Islam T et al (2024) Advancements and challenges in natural fiber-reinforced hybrid composites: a comprehensive review. SPE Polym 5(4):481–506
20. Mohammed M et al (2021) A review on the advancement of renewable natural fiber hybrid composites: prospects, challenges, and industrial applications. J Renew Mater 12(7)
21. Lubis S, Napitupulu FH, Abdullah I, Sitorus TB, Karim MA (2024) Study of Comparison heat energy disposal in hollow roof cooling water tank using natural material. J Adv Res Appl Sci Eng Technol 55(2):14–42
22. Ram Kishore S et al (2024) Natural fiber biocomposites via 4D printing technologies: A review of possibilities for agricultural bio-mulching and related sustainable applications. Prog Addit Manuf 9(1):37–67

23. Abdollahiparsa H, Shahmirzaloo A, Teuffel P, Blok R (2023) A review of recent developments in structural applications of natural fiber-reinforced composites (NFRCs). Compos Adv Mater 32:26349833221147540
24. Ali SA et al (2025) Review on the role of electrofuels in decarbonizing hard-to-abate transportation sectors: advances, challenges, and future directions. Energy Fuels
25. Thapliyal D et al (2023) Natural fibers composites: origin, importance, consumption pattern, and challenges. J Compos Sci 7(12):506
26. Singh A, Yadav BP (2024) Sustainable innovations and future prospects in construction material: a review on natural fiber-reinforced cement composites. Environ Sci Pollut Res 1–39
27. Ali SA et al (2025) State-of-the-art membrane solutions for direct air carbon capture (DACC): an overview on the current status and future directions. Energy Fuels 39(20):9285–9315
28. Tyagi A, Meena A, Bhandari M, Shahid M, Gupta BL, Meena SL (2024) A critical review on additive manufacturing of sustainable NFRCs: processing and applications. Green Mater 1–11
29. Nwodo MN, Anumba CJ (2019) A review of life cycle assessment of buildings using a systematic approach. Build Environ 162:106290
30. Islam S et al (2025) Thermoset and thermoplastic polymer composites reinforced with flax fiber: properties and application—a review. SPE Polym. 6(1):e10172
31. Prem Kumar R, Muthukrishnan M, Felix Sahayaraj A (2023) Effect of hybridization on natural fiber reinforced polymer composite materials–a review. Polym Compos 44(8):4459–4479
32. Tyagi A, Gupta BL, Mukhija L, Bhandari M, Meena A, Meena SL (2024) Mechanical and tribological behavior of NFRCs: a comprehensive analysis of properties and applications. Green Mater 1–12
33. Tasgin Y, Demircan G, Kandemir S, Acikgoz A (2024) Mechanical, wear and thermal properties of natural fiber-reinforced epoxy composite: cotton, sisal, coir and wool fibers. J Mater Sci 59(24):10844–10857
34. Repon MR, Islam T, Islam T, Alim MA (2024) Manufacture of polymer composites from plant fibers. Plant Biomass Deriv Mater Sourc Extr Appl 363–388
35. Saleh Alghamdi S, John S, Roy Choudhury N, Dutta NK (2021) Additive manufacturing of polymer materials: progress, promise and challenges. Polymers (Basel) 13(5):753
36. Prabhu R, Mendonca S, Bellairu PK, Dsouza RC, Bhat T (2024) Investigating the impact of TiO2 filler on abrasive wear characteristics of bamboo fiber-reinforced epoxy composites using the Taguchi method. World J Eng
37. Castro-Franco AD, Siqueiros-Hernández M, García-Angel V, Mendoza-Muñoz I, Vargas-Osuna LE, Magaña-Almaguer HD (2024) A review of natural fiber-reinforced composites for lower-limb prosthetic designs. Polymers (Basel) 16(9):1293
38. Sathish M, Radhika N, Venuvanka N, Rajeshkumar L (2024) A review on sustainable properties of plant fiber-reinforced polymer composites: characteristics and properties. Polym Int 73(11):887–943
39. Possidónio C, Farias AR, Domingos S, Cruz B, Luís S, Loureiro A (2025) Exploring supply-side barriers for commercialization of new biopolymer production technologies: a systematic review. Sustainability 17(3):820
40. Simões S (2024) High-performance advanced composites in multifunctional material design: state of the art, challenges, and future directions. Materials (Basel) 17(23):5997
41. Sathish S et al (2021) A review of natural fiber composites: extraction methods, chemical treatments and applications. Mater Today Proc 45:8017–8023
42. Navas E, Fernández R, Sepúlveda D, Armada M, Gonzalez-de-Santos P (2021) Soft grippers for automatic crop harvesting: a review. Sensors 21(8):2689
43. Harris F, Roby H, Dibb S (2016) Sustainable clothing: challenges, barriers and interventions for encouraging more sustainable consumer behaviour. Int J Consum Stud 40(3):309–318
44. Mylsamy B, Shanmugam SKM, Aruchamy K, Palanisamy S, Nagarajan R, Ayrilmis N (2024) A review on natural fiber composites: polymer matrices, fiber surface treatments, fabrication methods, properties, and applications. Polym Eng Sci 64(6):2345–2373
45. Banea MD, De Queiroz HFM, Neto HDEDE, Cavalcanti DKK Recent advances in development, characterization and joining of new sustainable materials. Mater Res Proc 41

Modelling Customer Satisfaction, Merchandise Returns and Repurchase Intention in Online Fashion Retailing

M. Krishnakumar, S. Mukthy, and P. Chithira

Abstract Returns significantly influence sales in the online fashion retail sector, often resulting from issues such as incorrect size, colour, style, or unmet expectations. High return rates are common in fashion e-commerce, with studies highlighting their prevalence. These returns impose substantial costs on retailers, including expenses related to return shipping, restocking, and potential revenue loss. This study explores factors such as website/app usability, information availability, pricing, delivery, and product quality to assess their impact on customer satisfaction in online fashion retail. It also examines how these factors influence product returns and repurchase intentions. Data were collected through a structured questionnaire administered to 200 respondents using a non-probability convenience sampling method. Structured Equation Modelling (SEM) was employed to analyse the relationships between variables. The results indicate that website/app usability and delivery significantly affect customer satisfaction, which in turn positively impacts repurchase intentions. Conversely, merchandise returns negatively influence repurchase intentions. Based on these findings, recommendations are made to enhance the customer experience and boost sales.

Keywords Customer satisfaction · Merchandise returns · Online retailing · Fashion · Repurchase intention

M. Krishnakumar (✉) · S. Mukthy · P. Chithira
Department of Fashion Management Studies, National Institute of Fashion Technology, Kannur, Kerala, India
e-mail: mkrishna.kumar@nift.ac.in

S. Mukthy
e-mail: Mukthy.s@nift.ac.in

S. S. Muthu (ed.), *Sustainable Solutions in Textiles and Fashion*, Sustainable Textiles: Production, Processing, Manufacturing & Chemistry,
https://doi.org/10.1007/978-3-032-10925-5_3

1 Introduction

Returns are a common aspect of online shopping, necessitating businesses to establish clear protocols for effective management [5]. In the Indian fashion sector, return rates are notably higher, with one out of every four items being returned. Research from Rest of World (2023) indicates that approximately 25–40% of online clothing purchases in India are returned, surpassing the return rates seen in the United States. These returns significantly impact sales in the online fashion retailing industry, often stemming from issues like incorrect sizing, colour discrepancies, style mismatches, or unmet customer expectations. The e-commerce landscape witnesses a considerable frequency of returns, which can be financially taxing for online retailers due to expenses related to return shipping, restocking, and potential revenue loss. Park and Kim [25] emphasize the importance of comprehending the interplay between various factors in online shopping and consumer purchase behaviour, with elements like security, information accessibility, shipping, quality, pricing, and time being closely linked to customer satisfaction [33]. Enhanced customer satisfaction not only fosters loyalty and retention but also boosts repeat sales significantly. Hence, there is a crucial need to devise strategies to minimize product returns and enhance customers' intent to repurchase, thereby elevating the overall customer experience and driving sales for online fashion brands. This study delves into these fundamental components, exploring factors such as website/app usability, information availability, pricing, delivery services, and product quality that influence customer satisfaction in online product purchases. Furthermore, the research investigates the intricate relationship between customer satisfaction, merchandise returns, and repurchase intentions in the realm of online fashion retailing in India. The study addresses the following research questions concerning customer satisfaction, merchandise returns, and repurchase intentions in the online fashion retailing landscape in India:

i. How do website/app usability, information availability, pricing, delivery services, and product quality impact customer satisfaction in online fashion retailing in India?
ii. In the context of online fashion retailing in India, how does customer satisfaction influence merchandise returns and repurchase intentions?
iii. How do merchandise returns affect repurchase intentions in the online fashion retailing sector in India?

2 Literature Review and Hypothesis Development

Several studies have delved into the realms of product returns [9, 11], customer satisfaction [3, 4, 6], and repurchase intentions [13]. Park and Kim [25] emphasize the significance of comprehending the relationship between various aspects of online shopping and consumer purchase behaviour within the online shopping context. Their research reveals that information satisfaction and relational benefits act as mediating factors influencing consumer shopping behaviour, with both factors playing a crucial role in consumers' commitment to an online store. Vasić et al. [33] highlight that multiple determinants such as security, information availability, shipping, quality, pricing, and time are interconnected with customer satisfaction, although the strength of their relationships may vary. In the Serbian market, shipping, pricing, and information availability wield a more substantial influence on customer satisfaction compared to quality, time, and security.

Hsu [16] developed a customer satisfaction index for online consumers modelled after the American Customer Satisfaction Index, proving to be an effective predictor of online consumer satisfaction and loyalty. Lin and Sun [21] categorized the factors influencing consumer satisfaction into external factors, including technology acceptance and website service quality, and internal factors encompassing hold up costs and other pertinent considerations.

2.1 *Website/App Usability*

Effective website and app usability play a critical role in online shopping as they directly shape the user experience and ultimately impact the success of a business. With advancements in e-commerce technologies and the widespread use of mobile devices and social media, companies now have the tools to enhance customers' shopping experiences and their interactions with brands, irrespective of time or location [7]. Key factors contributing to a seamless online customer experience encompass the ease of navigation and usage, perceived utility, enjoyable and practical features, personalization, social engagement, and compatibility across various devices [7]. In their research, [6] explore the correlation between website usability, consumer satisfaction, and the intention to use a website, as well as how satisfaction impacts usage intentions. Their findings underscore the significant impact of website usability on satisfaction levels, which, in turn, influences the intention to use the website. A positive outcome of an engaging online customer experience includes increased brand engagement, positive word-of-mouth recommendations, and repeat purchases. Hence, drawing from these insights, the first hypothesis of this study is formulated as:

> H1: Website/App Usability has a significant positive influence on Customer Satisfaction.

2.2 *Information Availability*

Karim [18] suggests that the information presented on websites/apps directly influences consumers' decision-making processes. Access to comprehensive product information allows consumers to understand the features, benefits, and potential drawbacks of a product, enabling them to make informed purchasing decisions.

The study reveals that information availability serves as a key motivating factor for online shopping [18]. Similarly, [33] identify information availability as a crucial variable with a substantial impact on e-customer satisfaction among various factors. The study underscores the significance of providing accurate, reliable, and complete information in online shopping experiences. Having easily accessible and comparable product information can help prevent returns resulting from incomplete or inaccurate details. Moreover, online shoppers can benefit from reading reviews and comments from other consumers before finalizing a purchase. Drawing insights from these studies, the second hypothesis of this research study is formulated as:

H2: Information Availability has a significant positive influence on Customer Satisfaction.

2.3 *Price*

Price holds a pivotal position in the realm of online shopping and significantly impacts customer satisfaction. In the online shopping landscape, customers are presented with a vast array of products from different retailers, allowing them to easily compare prices across various platforms. This accessibility makes customers more price-sensitive when considering the products they wish to buy. Rudansky-Kloppers [28] notes that customers typically anticipate online products to be priced lower than those found in physical stores. The pricing experience encountered by online buyers during their interactions with an online retailer can influence their decisions to either patronize the store again in the future or look elsewhere for their online shopping needs. Factors such as payment convenience, cost-effectiveness, and value for money play crucial roles in shaping market offerings for online businesses. Vasić et al. [33] identify pricing as one of the variables with the most significant impact among other factors in their study, aligning with other research indicating that product pricing can enhance customer satisfaction. According to [17], price exerts a notable influence on customer satisfaction, particularly during the initial stages of an online purchase. This implies that customers are more likely to feel satisfied if they perceive the pricing as fair and reasonable. Retailers must carefully craft their pricing strategies to attract and retain customers in a fiercely competitive market. Drawing from the aforementioned discussions, the third hypothesis of this study is posited as:

H3: Price has a significant positive influence on Customer Satisfaction.

2.4 *Delivery*

Multiple studies underscore the critical role of delivery services in the logistics of online purchases, with its performance directly influencing customer satisfaction levels [10, 14]. The delivery process in online transactions stands as a crucial determinant of customer satisfaction, impacting the choice to purchase from a specific online retailer. Any challenges experienced during the delivery process can have adverse effects on the overall purchase experience [10]. Handoko [14] investigates the correlation between product quality, delivery services, and customer satisfaction in the context of an online fashion retailer, highlighting that delivery services notably shape customer satisfaction. To enhance customer satisfaction, it is recommended that online retailers consistently provide reliable and secure delivery services that align with promised delivery timelines, fostering repeat purchases. Moreover, [36] reveal that delivery services, characterized by reliability, safety, and timeliness, directly impact e-satisfaction and e-trust, subsequently influencing e-loyalty. In essence, delivery elements such as appropriate packaging, delivery conditions, and punctuality are paramount in ensuring customer satisfaction and loyalty in the realm of online shopping. Drawing from these discussions, the fourth hypothesis of this study is formulated as:

H4: Delivery features has a significant positive influence on Customer Satisfaction.

2.5 *Quality*

The quality of a product is a crucial factor that directly influences customer satisfaction with their purchase and overall shopping experience. When shopping online, customers heavily rely on the product descriptions and images provided on the website to make informed purchasing decisions. If the actual quality of the product fails to meet expectations, customers are likely to experience dissatisfaction and may even opt to return the product. Quality service also plays a significant role in shaping customers' perceptions of a brand and their likelihood of engaging in future purchases. Ziaullah et al. [36] highlight that product quality and delivery services, encompassing factors like product variety, availability, reliable and safe delivery, and timeliness, have a direct impact on e-satisfaction and e-trust, subsequently influencing e-loyalty. Rita et al. [27] discover that three dimensions of e-service quality—website design, security/privacy, and fulfilment—significantly contribute to overall e-service quality, which in turn correlates with customer behaviour. Alex and Thomas [4] reveal that customer-perceived value is notably affected by product quality, service quality, and contextual experience. Moreover, a study by [11] aiming to evaluate the effect of e-service quality on customer satisfaction and repurchase intention among marketplace users indicates a significant impact of e-service quality on customer satisfaction and repurchase intentions. Drawing from the above discussions, the fifth hypothesis of this study is articulated as:

H5: Quality has a significant positive influence on Customer Satisfaction.

2.6 Customer Satisfaction

Customer satisfaction serves as a crucial metric reflecting consumers' contentment with a company's offerings, services, and capabilities, showcasing their overall perception of the value derived from a product or service. In the realm of online shopping, customer satisfaction extends beyond the purchased product or service and encompasses factors such as website navigation or usability [20], the online shopping experience [1], delivery times [3], return policies, among others [33]. Wang and Huarng [34] conducted a study on online customer satisfaction based on online reviews, revealing that receiving the desired product on time was a key factor contributing to favorable ratings, while customer service played a pivotal role in negative feedback. Satisfied customers highlighted timely receipt of requested items at a fair price without hidden charges as their primary concern, followed by quality customer assistance. Conversely, unsatisfied customers were primarily concerned about customer service via phone or email, product availability, and timely delivery of requested items.

Cattapan and Pongsakornrungsilp [8] explored the impact of channel integration on enhancing customer satisfaction, identifying six characteristics of retail channel integration—integrated product and price, promotion, transaction information, information access, and customer service—that influence customer satisfaction. The study also underscores the correlation between perceived risk, customer satisfaction, and purchase intention, positing that these factors contribute to increased customer satisfaction and product purchases. Miao [22] found that delivery service affects e-satisfaction, whereas customer service quality impacts trust and e-satisfaction, ultimately encouraging repurchases. Tzeng et al. [32] discovered that information quality, product quality, and cost savings enhance customer satisfaction in online holiday shopping, while product returns diminish customer satisfaction. They suggest that efficient after-sale services in handling product returns can smoothen the process and consequently boost customer satisfaction.

Additionally, this study explores post-purchase satisfaction through the lens of belief disconfirmation utilizing a theoretical framework of the expectancy disconfirmation theory [24]. Al-Jahwari et al. [2] highlight that product quality and service guarantees significantly influence satisfaction levels. Building on the insights from these discussions, the sixth and seventh hypotheses of this research study are formulated as:

H6: Customer Satisfaction has a significant negative influence on Merchandise returns

H7: Customer Satisfaction has a significant positive influence on Repurchase Intention.

2.7 *Merchandise Return*

Product returns in the realm of online shopping warrant meticulous study, as the actions taken by online retailers concerning returns can significantly impact customers' future purchasing behaviours. A research study by [5] reveals that online retailers aim to offer customers convenience in saving time and efficiently completing tasks. Respondents under time constraints particularly value flexible return policies and affordable, prompt delivery options. The findings indicate that individuals with limited time harbour negative perceptions of delayed deliveries and express uncertainties regarding the suitability of received products. They lean towards lenient return policies and appreciate the ability to return items hassle-free without incurring additional expenses.

Griffis et al. [13] explore the correlation between customers' product return experiences and subsequent shopping behaviours, noting that the return management process can positively influence customers' repurchase behaviours. Pei and Paswan [26] delve into consumers' online return behaviours, categorizing them as legitimate or opportunistic return behaviours. Their study reveals that legitimate return behaviours can foster increased customer loyalty, while opportunistic return behaviours may have adverse effects. Tzeng et al. [32] investigate how various factors impact customers' satisfaction during online shopping events, highlighting that product returns have a detrimental effect on customer satisfaction. Nevertheless, offering effective after-sales services can streamline the product return process and ultimately boost customer satisfaction. Building on the insights gleaned from these discussions, the eighth hypothesis of this study is proposed as:

H8: Merchandise Return has a significant negative influence on Repurchase Intention.

2.8 *Repurchase Intention*

When a customer expresses an intention to make future purchases from a specific brand or company, it signifies their satisfaction and trust in that brand. Various factors can influence a customer's repurchase intention, including product quality, customer service, pricing, brand reputation, and overall customer experience. A positive experience with a product or service, coupled with a perception of the brand's reliability and trustworthiness, increases the likelihood of customers repurchasing. Repeat purchases are crucial for online retailers as they are a reflection of customer satisfaction. Nguyen et al. [23] studied factors such as responsiveness, trust, convenience, delivery, information quality, and perceived website usability on customer satisfaction and repurchase intention, revealing that information quality, delivery, convenience, and perceived website usability have the most significant impacts, while responsiveness does not exhibit a significant influence.

Hellier et al. [15] developed a model outlining how seven essential elements impact consumer repurchase intention: service quality, equity and value, customer satisfaction, past loyalty, expected switching costs, and brand preference. The study findings suggest that while perceived quality does not directly enhance customer satisfaction, it does so indirectly through consumer equity and value perspectives. Past purchase loyalty is not directly linked to customer satisfaction, but brand preference acts as an intermediary factor between customer satisfaction and repurchase intention. Suleman et al. [29] identify trust and ease of use as significant factors influencing purchase decisions and repurchase intentions, with trust exerting a significant influence on consumer decisions and ease of use impacting repurchase intentions. Khoa et al. [19] explore the relationship between components contributing to customer relationships and the repurchase intention of fashion goods. Tandon et al. [30] discovered that website quality enhances customer satisfaction, thereby affecting repurchase intention.

3 Materials and Method

3.1 Research Model

The research model of this study is as shown in Fig. 1.

3.2 Research and Sampling Design

Quantitative research, primarily exploratory in nature, was utilized in this study, which aims to enhance customer satisfaction and comprehend the factors contributing to product returns in the online fashion industry. The study targeted customers across India who have made purchases from an online fashion brand. A convenient sampling method was employed to gather customer details. The study was executed through an online survey conducted via Google Forms. From the responses received, a sample size of 200 was identified out of the 225 surveys distributed.

3.3 Questionnaire Design

The questionnaire design encompassed a structured questionnaire comprising demographic information, occasion preferences, visit frequency, and willingness to recommend. The study's variables were measured using multiple items on a 5-point Likert scale, ranging from 1—Strongly Disagree to 5—Strongly Agree. The questionnaire was adapted from prior studies related to the variables under examination. A customer

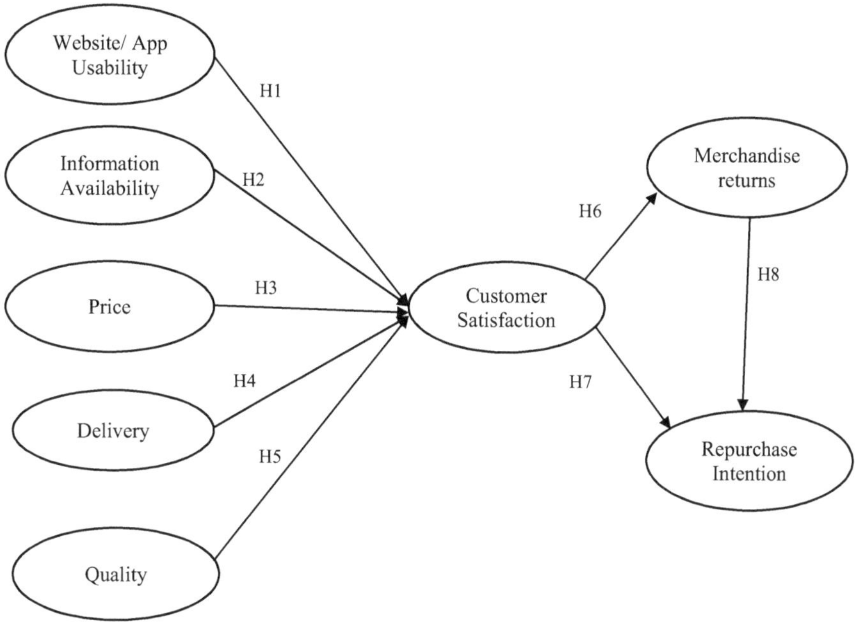

Fig. 1 Research model

survey via Google Forms was carried out among online fashion goods customers to collect insights on various challenges faced and comprehend their expectations.

3.4 *Measurement Items*

The measurement items utilized in this study were sourced from the references specified in Table 1.

4 Data Analysis

4.1 *Demographic Descriptive*

From Table 2, it is evident that out of the 200 respondents, 85% are female. The majority of respondents, 60%, fall within the 18–25 age group, with an equal percentage of 43.5% each for students and employed individuals. Additionally, 40.5% of the respondents belong to the income group categorized as not applicable. In terms of occasion preference, 47.5% of respondents favor Casual attire, while 26%

Table 1 Scales adopted

Variable		Item Scale (Likert–5 Point)	References
Website/app availability	1.1	Using this website/app has improved the quality of the online shopping experience of fashion products	[35]
	1.2	Conducting a transaction is easy on the website/app	
	1.3	The website/app is easy to navigate	
Information availability	2.1	The website/app is easy to navigate	[23]
	2.2	The product images provided are accurate to the product received	
	2.3	The customer reviews and ratings provided are important in making the purchase decision	
Price	3.1	I am satisfied with the price of the product purchased	[32]
	3.2	I am satisfied with the offers and pricing provided on the website/app compared to other shopping sites	
	3.3	Sufficient information about the offers and any additional fees or charges are provided on the website/app	
Delivery	4.1	The product had the right package while delivered	[23]
	4.2	The product was delivered on time	
	4.3	Condition of the product upon arrival was good	
Quality	5.1	The quality of products sold are high	[32]
	5.2	The website/app delivers orders or services in a timely manner	
	5.3	Provides easy returns and refunds on the website/app	
Customer satisfaction	6.1	I am satisfied with the online payment system security of the website/app	[35]
	6.2	I am satisfied with the online shopping experience on the website/app	
	6.3	I like the content and layout of the website/app	
Merchandise return	7.1	I return the product received if there is any defect	[26]
	7.2	I return the wrong product received because of the seller's fault	
	7.3	I return the product because of changed mind	

(continued)

Table 1 (continued)

Variable		Item Scale (Likert–5 Point)	References
	7.4	I return the product because I found it elsewhere for lower price	
Repurchase intention	8.1	I intend to shop at this website/app again in future	[9]
	8.2	I am likely to visit this website/app again in future	
	8.3	I will shop more at this website/app in the future	

opt for Everyday Essentials, 12% for Party wear, 9% for Festive attire, and 5.5% for Formal attire. Regarding the frequency of visits, 44.5% of respondents visit Sometimes, 26% visit Often, 24.5% visit Rarely, 4% visit Always, and 1% visit Never. As for willingness to recommend, 44% of respondents are Likely to recommend, 30% are Neutral, 22.5% are Very Likely, 2% are Unlikely, and 1.5% are Very Unlikely.

4.2 Reliability Analysis

Cronbach's alpha was employed to assess the internal consistency (reliability) of the variables. It is a widely used method when multiple Likert scale statements collectively form a construct, helping to ascertain the reliability of those items in measuring the given construct. As per the table, all variables exhibit Cronbach's alpha values exceeding 0.7, signifying a strong level of internal consistency for the utilized construct (Table 3).

4.3 Discriminant Validity

Discriminant validity was evaluated through the Fornell and Larcker [12] criterion, which involved comparing the square root of each Average Variance Extracted (AVE) on the diagonal with the correlation coefficients between constructs (off-diagonal) in the corresponding rows and columns. The analysis confirms the presence of discriminant validity in this measurement model, indicating clear distinction between the constructs (Table 4).

Table 2 Demographic descriptive

Demography	Details	No. of respondents	Percentage (%)
Gender	Male	30	15
	Female	170	85
Age	18–25 years	120	60
	26–35 years	74	37
	36–45 years	4	2
	Above 46 years	2	1
Occupation	Student	87	43.5
	Self-employed	16	8.0
	Employed	87	43.5
	Unemployed	10	5.0
Income	Less than 20,000 INR	43	21.5
	Between 20,000 – 40,000 INR	39	19.5
	Between 40,000 – 60,000 INR	15	7.5
	Above 60,000 INR	22	11.0
	Not applicable	81	40.5
Occasion Preference	Casual	95	47.5
	Party	24	12.0
	Formal	11	5.5
	Festive	18	9.0
	Everyday Essentials	52	26.0
Frequency of visit	Always	8	4.0
	Often	52	26.0
	Sometimes	89	44.5
	Rarely	49	24.5
	Never	2	1.0
Willingness to recommend	Very Likely	45	22.5
	Likely	88	44.0
	Neutral	60	30.0
	Unlikely	4	2.0
	Very Unlikely	3	1.5

Source Primary data

Table 3 Construct reliability and validity

	Cronbach's alpha	Composite reliability (rho_a)	Average variance extracted (AVE)
Acceptable range	≥ 0.7 (0 to 1)	≥ 0.7 (0 to 1)	≥ 0.5
Customer Satisfaction (CS)	0.773	0.788	0.689
Delivery (D)	0.798	0.826	0.715
Information Availability (IA)	0.738	0.749	0.658
Price (P)	0.788	0.787	0.706
Quality (Q)	0.704	0.714	0.626
Repurchase Intention (RI)	0.909	0.915	0.846
Website/App Usability (WAU)	0.797	0.798	0.711

Source Primary data

4.4 *Structural Equation Modelling*

Structural Equation Modelling (SEM) is a sophisticated statistical analysis technique employed to examine structural relationships among variables. Integrating factor analysis and multiple regression analysis, SEM is utilized to analyse the structural associations between observed variables and latent constructs. Researchers favour this method as it enables the estimation of multiple interrelated dependencies within a single analysis. The outcomes derived from the data in SmartPLS can be summarized as follows.

4.4.1 Measurement Model

The diagram illustrates the variables Website/App Usability (WAU), Information Availability (IA), Price (P), Delivery (D), Quality (Q), Customer Satisfaction (CS), Merchandise Returns (MR), and Repurchase Intention (RI). With Cronbach's alpha exceeding 0.4 and Average Variance Extracted (AVE) surpassing 0.5, the data demonstrates reliability and validity. According to [31], a factor loading above 0.30 typically signifies a moderate correlation between the variable and the measurement items.

Table 4 Discriminant validity test—cross Loading

Discriminant validity test—cross Loading								
	CS	D	IA	MR	P	Q	RI	WAU
1.1	0.652	0.534	0.559	0.19	0.461	0.528	0.477	**0.829**
1.2	0.674	0.459	0.535	0.216	0.453	0.518	0.458	**0.86**
1.3	0.687	0.519	0.523	0.283	0.512	0.558	0.505	**0.841**
2.1	0.553	0.44	**0.844**	0.076	0.556	0.498	0.428	0.529
2.2	0.577	0.496	**0.856**	0.122	0.6	0.61	0.482	0.531
2.3	0.477	0.486	**0.728**	0.372	0.353	0.439	0.403	0.495
3.1	0.549	0.422	0.676	0.157	**0.753**	0.543	0.443	0.536
3.2	0.56	0.475	0.464	0.116	**0.865**	0.596	0.491	0.46
3.3	0.521	0.365	0.435	0.079	**0.895**	0.53	0.491	0.418
4.1	0.403	**0.737**	0.447	0.144	0.454	0.674	0.384	0.395
4.2	0.554	**0.915**	0.508	0.296	0.466	0.639	0.522	0.588
4.3	0.532	**0.874**	0.52	0.254	0.373	0.621	0.5	0.513
5.1	0.612	0.486	0.554	0.101	0.537	**0.785**	0.44	0.608
5.2	0.438	0.685	0.485	0.175	0.509	**0.749**	0.367	0.406
5.3	0.565	0.651	0.475	0.126	0.532	**0.836**	0.466	0.464
6.1	**0.747**	0.358	0.49	0.094	0.398	0.456	0.433	0.658
6.2	**0.885**	0.519	0.568	0.181	0.643	0.593	0.606	0.684
6.3	**0.853**	0.581	0.588	0.248	0.556	0.659	0.551	0.647
7.1	0.215	0.281	0.222	**1**	0.141	0.163	0.102	0.273
8.1	0.603	0.558	0.522	0.137	0.526	0.541	**0.906**	0.53
8.2	0.527	0.494	0.474	0.108	0.474	0.476	**0.937**	0.5
8.3	0.634	0.489	0.491	0.04	0.556	0.473	**0.916**	0.536

Source Primary data
The values above 0.7 are shown in bold

4.4.2 Structural Model

Among the variables, the path coefficients for Website/App usability and customer satisfaction ($\gamma = 0.509$, $t = 4.709$, $p < 0.05$), price and customer satisfaction ($\gamma = 0.184$, $t = 2.478$, $p < 0.05$), Quality and customer satisfaction ($\gamma = 0.254$, $t = 2.274$, $p < 0.05$), as well as customer satisfaction and repurchase intention ($\gamma = 0.614$, $t = 9.380$, $p < 0.05$) were positive and statistically significant. This outcome supports hypotheses H1, H3, H5, and H7.

The path coefficient for the variables information availability and customer satisfaction ($\gamma = 0.139$, $t = 1.413$, $p > 0.05$) was positive but not statistically significant, thereby partially supporting hypothesis H2. Similarly, the relationship between customer satisfaction and merchandise returns ($\gamma = -0.263$, $t = 1.051$, $p > 0.05$) was negative but not significant, partially supporting hypothesis H6.

Table 5 Measurement model

Constructs	Items	Loadings	Cronbach's alpha	Composite reliability	AVE
Website/app availability	1.1	0.829	0.797	0.798	0.711
	1.2	0.860			
	1.3	0.841			
Information availability	2.1	0.844	0.738	0.749	0.658
	2.2	0.856			
	2.3	0.728			
Price	3.1	0.753	0.788	0.787	0.706
	3.2	0.865			
	3.3	0.895			
Delivery	4.1	0.737	0.798	0.826	0.715
	4.2	0.915			
	4.3	0.874			
Quality	5.1	0.785	0.704	0.714	0.626
	5.2	0.749			
	5.3	0.836			
Customer satisfaction	6.1	0.747	0.773	0.788	0.689
	6.2	0.885			
	6.3	0.853			
Merchandise return	7.1	0.821	0.701	0.710	0.616
	7.2	0.434			
	7.3	0.671			
	7.4	0.567			
Repurchase intention	8.1	0.906	0.909	0.915	0.846
	8.2	0.937			
	8.3	0.916			

Source Primary data

On the other hand, the path coefficients for variables delivery and customer satisfaction ($\gamma = -0.109$, $t = 1.181$, $p > 0.05$) as well as merchandise returns and repurchase intention ($\gamma = 0.032$, $t = 0.811$, $p > 0.05$) were not statistically significant. These results do not provide support for hypotheses H4 and H8. The R^2 values for customer satisfaction, merchandise return, and repurchase intention factors are 0.764, 0.069, and 0.368, respectively (Fig. 2, Table 6).

Table 6 Structural Equation Model – Direct Effect

Path	Beta	SD	T-stat	p-value	Hypo thesis	Decision
Website/app usability > customer satisfaction	0.509	0.108	4.709	0.000	H1	Supported
Information availability > customer satisfaction	0.139	0.098	1.413	0.158	H2	Partially supported
Price > customer satisfaction	0.184	0.074	2.478	0.013	H3	Supported
Delivery > customer satisfaction	-0.109	0.092	1.181	0.237	H4	Not supported
Quality > customer satisfaction	0.254	0.112	2.274	0.023	H5	Supported
Customer satisfaction > merchandise returns	-0.263	0.250	1.051	0.293	H6	Partially Supported
Customer satisfaction > repurchase intention	0.614	0.066	9.380	0.000	H7	Supported
Merchandise returns > repurchase intention	0.032	0.135	0.239	0.811	H8	Not supported
R^2 value						
Customer satisfaction	0.764					
Merchandise returns	0.069					
Repurchase intention	0.368					

Source Primary data

4.4.3 SEM Indirect Effect

Table 7 presents the indirect impact of the variables that hold significance in the research model, accompanied by their standard deviation (SD), t-statistics, and p-values.

The analysis of Table 8 indicates that the following indirect effects hold significance in the paths: i) Website/App Usability > Customer Satisfaction > Repurchase intention ii) Price > Customer Satisfaction > Repurchase intention iii) Quality > Customer Satisfaction > Repurchase intention. These findings highlight the significant roles played by Website Usability, Price, Quality, and Customer Satisfaction in influencing repurchase intention in this study.

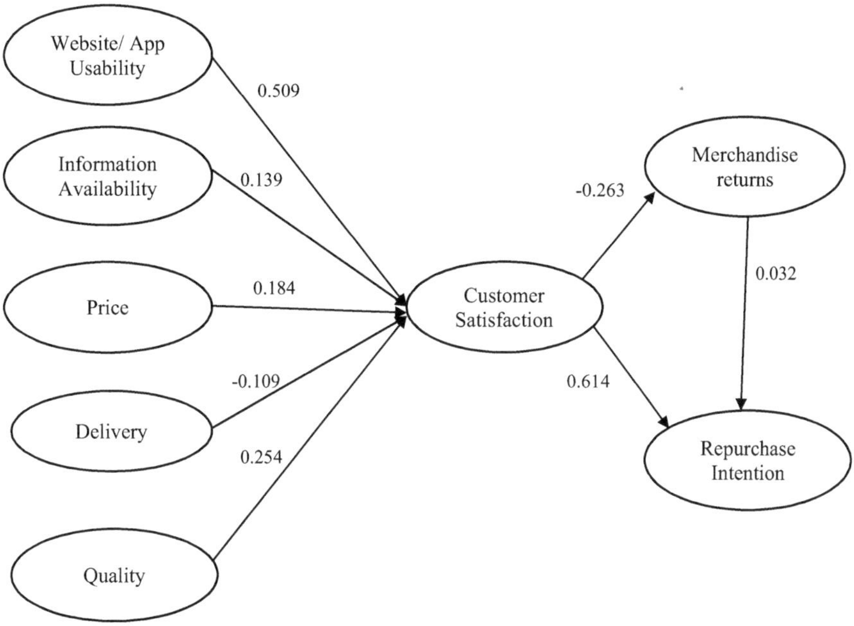

Fig. 2 Structural equation modelling. *Source* Primary Data and developed by author

Table 7 Structural equation model—indirect effect

Indirect effect	Estimate coefficient	SD	T stat	P-value	Decision
WU > CS > RI	0.159	0.065	2.45	0.014	Significant
P > CS > RI	0.072	0.032	2.258	0.024	Significant
Q > CS > RI	0.178	0.055	3.255	0.001	Significant

Source Primary data

4.4.4 Model Fit

The goodness of fit of a statistical model assesses its alignment with a given set of observations. It quantifies the disparity between observed values and those anticipated by the model.

SRMR serves as an absolute measure of fit, where a value of zero signifies a perfect fit without penalizing model complexity. Typically, a value below 0.08 is indicative of a good fit. In this case, the SRMR value of 0.078 falls below 0.08, affirming that the model fits well.

Table 8 Model fit

Model fit test			
		Saturated model	Estimated model
Fitness test	Acceptable range for SRMR	≤ 0.08	≤ 0.08
Goodness of fit	SRMR	0.078	0.08
Exact fit	d_ULS	1.581	2.060
Exact fit	d_G	0.792	0.831
(N-1)*L	Chi-square	878.046	911.902
Incremental fit measure	NFI	0.704	0.701

Source Primary data

5 Discussions

The study aimed to investigate the factors influencing customer satisfaction and understand their impact on product returns and repurchase intention in the online fashion industry. Key findings from the study are as follows:

Website/App Usability demonstrates a positive and highly significant impact on Customer Satisfaction. This aligns with the research by [6], indicating that website usability affects satisfaction, leading to higher intention to use. The significance of Website/App Usability exceeds that of other online variables tested, emphasizing the importance of enhancing every aspect related to it. Ensuring a consistent user experience between the website and app, offering clear navigation with a user-friendly interface, and ensuring fast loading times across various devices are crucial aspects to focus on for improvement.

Information availability also shows a significant influence on Customer Satisfaction. This finding is consistent with the research of [18] and [33], underscoring information availability as a crucial factor impacting e-customer satisfaction. To enhance customer satisfaction, e-Marketers should consistently provide clear and reliable information on their websites and other media platforms to keep consumers informed about products and offers. Leveraging social media can effectively disseminate information to consumers. Utilizing data analytics, machine learning, and artificial intelligence can enhance personalization and recommendations. Implementing personalized product recommendations based on customer data can aid in styling choices, potentially increasing customer purchases. Tailoring recommendations to individual customers based on their past purchases and browsing behaviour can improve customer satisfaction and encourage repeat purchases.

It is observed that Price positively influences Customer Satisfaction, aligning with findings from [33, 28, 17]. Pricing strategies can be tailored based on competitor analysis, with the study's notable differences possibly reflecting varying price sensitivities among customer segments. Marketers can optimize pricing tactics by setting different price points for diverse groups and customizing promotional efforts to enhance efficacy. Offering added incentives like discounts, coupons, or free shipping can incentivize additional purchases and enhance overall customer experiences.

On the other hand, Delivery does not exhibit a significant impact on Customer Satisfaction, contrary to research by [15–36] that highlight the direct influence of delivery services on e-satisfaction. This disparity could be attributed to the more aesthetic-focused nature of fashion products, which may not be as time-sensitive as other goods. Consequently, the delivery of fashion items may not significantly affect customer satisfaction.

Quality, however, shows a positive effect on Customer Satisfaction, consistent with the findings of [36, 27] emphasizing the direct impact of product quality on e-satisfaction and customer behaviour. Issues related to quality, such as garment colour or fit discrepancies, can be addressed by providing high-quality images and videos showcasing product details and dimensions. Ensuring consistent sizing based on customer feedback can also enhance customer satisfaction.

Additionally, the study reveals that Customer Satisfaction negatively influences merchandise returns, aligning with research by [32, 13, 26] indicating the adverse impact of customer satisfaction on returns. For fashion products, customer satisfaction primarily centres on aesthetic features and item fit, leading to lower return rates when customers are highly satisfied with the design, style, or colour. Unless significant fit issues arise, customers are less likely to return fashion items they genuinely appreciate.

It has been observed that merchandise returns do not impact repurchase intention significantly. As mentioned earlier, the primary factors influencing repurchase decisions for fashion products revolve around their design, style, and colour. Customers who are drawn to a specific design, style, or colour are more likely to repurchase from the same online platform. While it may be challenging to completely eliminate merchandise returns in a fashion marketplace, measures can be taken to reduce their frequency. Providing detailed and accurate information, multiple images from various angles, videos, and virtual try-on options can help in this regard. Despite these efforts, returns may still occur, underscoring the importance of offering hassle-free return processes to maintain customer satisfaction. Regular collection of customer feedback through surveys and reviews is crucial for identifying areas needing improvement and implementing necessary changes across all stages of the online shopping journey. Customers can also be incentivized to provide feedback, fostering a culture of responsiveness and engagement.

Furthermore, the study findings indicate that customer satisfaction significantly influences repurchase intention. This aligns with research by [19, 30], highlighting the impact of website quality on repurchase intention through enhanced customer satisfaction. Online retailers can leverage technology solutions like virtual try-on and augmented reality to offer customers a more accurate depiction of how garments will fit and look, thereby reducing returns stemming from sizing or style discrepancies. It is essential to provide product ratings and reviews on both the website and the application, as these play a vital role in helping customers make informed purchase decisions. In addition to ratings, including reviews accompanied by images or videos can offer customers verified opinions from other users, fostering trust and influencing purchase decisions.

6 Practical Implications for Asian Business

Merchandise returns pose a common challenge for online fashion retailers in Asia. To address this issue, retailers can deploy various strategies such as detailed product descriptions, sizing charts, free returns and exchanges, enhanced customer experiences, and technological solutions for the Asian consumers. By implementing these strategies, online retailers in Asia can minimize return rates, leading to increased customer satisfaction, loyalty, and repeat business.

To minimize product returns, Asian retail companies should address the root causes behind the returns. Asian customers prefer a return process that is uncomplicated, clear, and seamless. This may involve actions like enhancing product descriptions and images for accurate representation, providing detailed size charts, addressing quality control issues, among other factors focusing on Asian consumers. When customers initiate a return, it often signifies dissatisfaction with their purchase, and the handling of returns plays a significant role in improving their experience. A straightforward and hassle-free return process, coupled with clear communication, can enhance customer satisfaction by showing that the company values feedback and is committed to rectifying any issues. Conversely, a complex return procedure that demands excessive effort from customers or lacks supportive customer service can leave a negative impression. A smooth return experience has the potential to increase the likelihood of customers repurchasing from the company in the future, as it demonstrates reliability and a focus on customer well-being. Conversely, a cumbersome return process may deter customers from returning to the company and may even dissuade others from engaging with the brand. In Asian markets, customer satisfaction plays a crucial role in shaping the link between product returns and repurchase intentions in various ways.

7 Conclusion

The study aims to explore the impact of several variables in online retailing, including website usability, information availability, price, quality, and delivery, on customer satisfaction, merchandise returns, and repurchase intention. With the exception of delivery and merchandise returns, all variables show varying degrees of influence on customer satisfaction and repurchase intention. The significant roles played by Website Usability, Price, Quality, and Customer Satisfaction in Repurchase Intention highlight the importance of focusing on these aspects to enhance customer satisfaction and drive repeat purchases. Furthermore, improving merchandise return policies and communication regarding returns can further contribute to overall customer satisfaction. The proposed model offers a foundational framework for understanding the dynamics of website usability, information availability, pricing, quality, delivery, customer satisfaction, merchandise returns, and repurchase intention within the online fashion retail sector in India. The study's insights hold managerial implications for online fashion retailers in India and offer valuable research implications for scholars in the field, aiding retailers in understanding the influence of various factors on customer satisfaction, merchandise returns, and repurchase intention to shape effective market segmentation and promotional strategies. Additionally, this paper contributes to the existing literature on customer satisfaction, merchandise returns, and repurchase intention in the online fashion retail industry in India and Asian context, providing valuable insights for research scholars seeking deeper understanding in this domain.

7.1 Limitations and Future Scope for Research

Limitations and future research directions: This study primarily focuses on merchandise returns and repurchase intention in online fashion retailing among Indian consumers, therefore, caution should be exercised when generalizing and applying the findings to other geographical and demographic settings. Future studies may explore different geographical, demographic, and socio-cultural environments as per their specific requirements. The proposed framework may also be adapted for use in other online retail sectors such as consumer durables, books, electronics, etc. Additionally, factors like website layout, checkout options, and payment methods could also influence customer satisfaction, merchandise returns, and repurchase intention, suggesting avenues for further investigation in future research endeavours.

References

1. Abdeldayem MM (2010) A study of customer satisfaction with online shopping: evidence from the UAE. Int J Adv Media Commun 4(3):235–257
2. Al-Jahwari NS, Khan MF, Kalbani GK, Khansouri SS (2018) Factors influencing customer satisfaction of online shopping in oman–youth perspective. Hum Soc Sci Rev 6:64–73
3. Alam SS, Ali MH, Omar NA, Hussain WM (2020) Customer satisfaction in online shopping in growing markets: an empirical study. Int J Asian Bus Inform Manag 11(1):78–91
4. Alex D, Thomas S (2011) Impact of product quality, service quality and contextual experience on customer perceived value and future buying intentions. Euro J Bus Manag 3(3):307–315
5. Alreck PL, DiBartolo GR, Diriker M, Dover HF, Passyn KA, Settle RB (2009) Time pressure, time saving and online shopping: exploring a contradiction. J Appl Bus Res 25(5):85–92
6. Belanche D, Casalo LV, Guinalı́u M (2012) Website usability, consumer satisfaction and the intention to use a website: the moderating effect of perceived risk. J Retail Consum Serv 19(1):124–132
7. Bilgihan A, Kandampully J, Zhang T (2016) Towards a unified customer experience in online shopping environments Antecedents and outcomes. Int J Qual Serv Sci 8(1):102–119

8. Cattapan T, Pongsakornrungsilp S (2022) Impact of omnichannel integration on Millennials' purchase intention for fashion retailer. Cogent Bus Manag 9(1):1–26
9. Chatzoglou P, Chatzoudes D, Savvidou A, Fotiadis T, Delias P (2022) Factors affecting repurchase intentions in retail shopping: an empirical study. Heliyon 8(9):1–18
10. Cihan C, Krisztina P, Akos V (2017) Try not to be late!–The Importance of delivery service in online shopping. Organiz Mark Emerg Econ 8(16):177–192
11. Fared MA, Darmawan D, Khairi M (2021) Contribution of e-service quality to repurchase intention with mediation of customer satisfaction: Study of online shopping through marketplace. J Market Bus Res 1(2):93–106
12. Fornell C, Larcker DF (1981) Evaluating structural equation models with unobservable variables and measurement error. J Mark Res 18(1):39–50. https://doi.org/10.2307/3151312
13. Griffis SE, Rao S, Goldsby TJ, Niranjan TT (2012) The customer consequences of returns in online retailing: an empirical analysis. J Oper Manag 30(4):282–294
14. Handoko LP (2016) The effect of product quality and delivery service on online customer satisfaction in Zalora Indonesia. J EMBA 4(1):1189–1199
15. Hellier PK, Geursen GM, Carr RA, Rickard JA (2003) Customer repurchase intention: a general structural equation model. Eur J Mark 37(11):1762–1800
16. Hsu S-H (2008) Developing an index for online customer satisfaction: adaptation of American customer satisfaction index. Expert Syst Appl 34:3033–3042
17. Jiang P, Rosenbloom B (2005) Customer intention to return online: price perception, attribute-level performance, and satisfaction unfolding over time. Eur J Mark 39(1/2):150–174
18. Karim RA (2013) Customer satisfaction in online shopping: a study into the reasons for motivations and inhibitions. IOSR J Bus Manag 11:13–20
19. Khoa BT, Nguyen TD, Nguyen VT-T (2020) Factors affecting customer relationship and the repurchase intention of designed fashion products. J Distrib Sci 18(2):17–28
20. Lee K, Joshi K (2006) Development of an integrated model of customer satisfaction with online shopping. SIGMIS-CPR'06 53–56
21. Lin GT, Sun CC (2009) Factors influencing satisfaction and loyalty in online shopping: an integrated model. Online Inform Rev 33(3):1468–4527, 458–475
22. Miao M, Jalees T, Zaman SI, Khan S, Khan S, Javed MK (2021) The influence of e-customer satisfaction, e-trust and perceived value on consumer's repurchase intention in B2C e-commerce segment. Asia Pac J Mark Logist 34(10):2184–2206
23. Nguyen L, Nguyen TH, Tan TK (2021) An empirical study of customers' satisfaction and repurchase intention on online shopping in Vietnam. J Asian Finan Econ Bus 8(1):971–983
24. Oliver RL (1977) Effect of expectation and disconfirmation on postexposure product evaluations: an alternative interpretation. J Appl Psychol 62(4):480–486
25. Park C-H, Kim Y-G (2003) Identifying key factors affecting consumer purchase behavior in an online shopping context. Int J Retail Distrib Manag 31(1):16–29
26. Pei Z, Paswan A (2018) Consumers' legitimate and opportunistic product return behaviors in online shopping. J Electron Commer Res 19(4):301–319
27. Rita P, Oliveira T, Farisa A (2019) The impact of e-service quality and customer satisfaction on customer behavior in online shopping. Heliyon 5:e02690, ScienceDirect, 1–14
28. Rudansky-Kloppers S (2014) Investigating factors influencing customer online buying satisfaction in Gauteng, South Africa. Int Bus Econ Res J 13(5):1187–1198
29. Suleman D, Sabil S, Rusiyati S, Sari I, Rachmawati S, Nurhayaty E, Parancika RB (2021) Exploring the relationship between trust, ease of use after purchase and switching re-purchase intention. Int J Data Netw Sci 5:465–470
30. Tandon U, Kiran R, Sah AN (2017) Customer satisfaction as mediator between website service quality and repurchase intention: an emerging economy case. Serv Sci 9(2):106–120
31. Tavakol M, Wetzel A (2020) Factor analysis: a means for theory and instrument development in support of construct validity. Int J Med Educ 6(11):245–247. https://doi.org/10.5116/ijme.5f96.0f4a
32. Tzeng S-Y, Ertz M, Jo M-S, Sarigollu E (2020) Factors affecting customer satisfaction on online shopping holiday. Mark Intell Plan 39(4):516–532

33. Vasić N, Kilibarda M, Kaurin T (2018) The influence of online shopping determinants on customer satisfaction in the Serbian market. J Theor Appl Electron Commer Res 4(2):70–89
34. Wang M, Huarng AS (2002) An empirical study of internet store customer post-shopping satisfaction. IACIS 632–638
35. Yo PW, Kee DM, Yu JW, Hu MK, Jong YC, Ahmed Z, Nair RK (2021) The influencing factors of customer satisfaction: a case study of shopee in Malaysia. Stud Appl Econ 39(12):1–16
36. Ziaullah M, Feng Y, Akhter SN (2014) E-loyalty: The influence of product quality and delivery services on e-trust and e-satisfaction in China. Int J Adv Res Technol 3(10):20–31

Revolutionising Sustainable Fashion: The Role of Nanotechnology

R. Surjit, J. G. Haritha, and G. Tejaswini

Abstract The textile industry is one of the largest polluters in the world. The production of raw materials, spinning them into fibres or yarns, weaving or knitting fabrics, dyeing, printing and finishing requires enormous water and chemicals, including pesticides for growing cellulosic fibres like cotton. Carbon dioxide emissions from these processes leave a significant carbon footprint. Consumers further contribute to environmental damage through water and chemicals used in washing and the release of microplastics. Less than half of the used clothes are collected for reuse or recycling, and only 1% is recycled into new garments, leading to excessive textile waste. A sustainable future requires alternatives to tackle these problems. Nanotechnology, the production of materials at the nanoscale (1–100 nm or 10^{-9} m), offers innovative solutions for sustainability in fashion. It enhances fabric properties, improves durability and reduces environmental impact. Nanocoating using zinc oxide and titanium oxide provides UV protection against microbial properties and self-cleaning abilities, reducing the need for excessive chemical treatments. Recent advancements include nano-enhanced biodegradable fibres, waterless dyeing techniques and moisture-wicking smart textiles contributing to resource efficiency and circular fashion. Nano-cellulose-based textiles replace synthetic fibres with biodegradable materials, reducing dependency on petroleum-based materials. Western Europe was the largest market for nanotech clothing in 2021. The global nanotechnology market was valued at $3.69 billion in 2022. It is expected to grow at a CAGR of 33. 1% from 2023 to 2030 (Grand View Research, Nanotechnology market size, share & trends analysis report [Online] https://www.grandviewresearch.com/industry-analysis/nanotechnology-market-report. Accessed on 23rd May 2025, 2023). The expanding use of nanotechnology in apparel in developing countries is fueling its rapid growth and future potential. This chapter will discuss the nanomaterials usage in textiles, energy-efficient nano manufacturing processes, different nano finishes, scalability challenges and the future of nanotechnology in aiding sustainable fashion. As the industry moves towards a responsible approach, nanotechnology helps in a transformative path to sustainable fashion.

R. Surjit (✉) · J. G. Haritha · G. Tejaswini
Department of Fashion Technology, PSG College of Technology, Coimbatore, Tamil Nadu, India
e-mail: rst.fashion@psgtech.ac.in

S. S. Muthu (ed.), *Sustainable Solutions in Textiles and Fashion*, Sustainable Textiles: Production, Processing, Manufacturing & Chemistry,
https://doi.org/10.1007/978-3-032-10925-5_4

Keywords Sustainability · Nanotechnology · Durability · Nano finishes · Nanotextiles

1 Introduction

In recent years, the fashion industry has been facing pressure to adapt itself to sustainable practices that address environmental concerns without compromising aesthetics and functionalities. Nanotechnology plays a transformative role in advancing sustainable fashion by introducing innovative methods that reduce environmental stress and promote resource efficiency, acting as a bridge between performance and sustainability. Nanotechnology introduces new possibilities across various stages of textile production, from fibres to finishing by manipulating materials at the nanoscale. This nanoscale integration helped in the development of fabric with water repellency, antibacterial properties and enhanced durability of fabrics. Nanotechnology helps to lower the water usage and discharge of toxic effluents, supporting eco-friendly production methods. Biomimetic innovations inspired by natural phenomena and models, and their replication on fabric handled at the nanoscale, further contribute to the sustainably functional designs. These developments show just the beginning of how nanotechnology is reshaping fashion with sustainability as its core. The global nanotechnology market was valued at $3.69 billion in 2022. It is expected to grow at a CAGR of 33.1% from 2023 to 2030 [1]. The expanding use of nanotechnology in apparel in developing countries is fueling its rapid growth and future potential. This chapter will discuss the nanomaterials usage in textiles, energy-efficient nano manufacturing processes, different nano finishes, scalability challenges and the future of nanotechnology in aiding sustainable fashion.

2 Nanomaterials in Textiles

Nanomaterials are incredibly small particles with at least one of their dimensions ranging from 1 to 100 nm. Based on its dimension, nanomaterials are classified into 0, 1, 2 and 3D nanomaterials.

- The 0-dimensional structures include particles in which all of their dimensions are less than 100 nm. Example: Nanoparticles, quantum dots.
- The 1-dimensional nanomaterials have a thickness or diameter on the nanoscale. Example: nanowires, nanotubes.
- 2D nanostructures are mostly sheets, thin films or coatings with 1 of their dimensions (thickness)in the nanoscale. Example: Graphene sheets.
- 3D nanomaterials are not in the nanoscale at any dimension and are made of nanomaterials such as reinforcing nanoparticles or nanofibres in polymer matrix, arranging nanotubes or wires as sheets and stacked nano-layered materials [2].

Nanomaterials exhibit unique properties such as high strength, high reactivity, and high surface area due to their small size and are widely used in industries including electrical, medicine, energy, environmental science, fashion and textile industries. In the fashion and textile industries, nanomaterials help in enhancing fabric properties, adding new functionalities, ensuring better performance and strength without compromising the comfort, quality and appearance of fabric or textile material. The utilisation of the nanomaterials in textile and fashion supports to development of long-lasting, smart and sustainable clothing that aligns with current trends and modern lifestyle needs. The upcoming section explores different nanomaterials used in these advancements, including nanoparticles, nanofibres, nanotubes, nanocomposites and other related substances.

2.1 Nanoparticles

Nanoparticles are incredibly small particles; all of their dimensions are less than 100 nm. These particles can be manufactured either by a top-down (bigger to smaller particles) or bottom-up approach (building the particles by chemical process). The nanoparticle plays an important role in various sectors, including the textile and fashion industry, directing the textile industry to sustainability and reshaping future fashion trends. Nanoparticles such as carbon, titanium dioxide, magnesium oxide, silver nanoparticles, etc., are incorporated in fibres, yarns, finishing, dyeing and different stages of the fabric manufacturing process, which are discussed below in detail.

2.2 Nanofibres

Nanofibres are extremely thin filaments having its a length in micrometres. As the diameter of the nanofibres decreases, the strength of the nanofibres increases [3]. Nanofibres are produced by electrospinning methods, even though there are some other methods, electrospinning is very popular and widely adopted. The functions of nanofibres are improved by incorporating nanoparticles such as TiO_2 during electrospinning, or by coating and spraying. Nanofibres have very high flexibility, breathability, high surface area, making them suitable for many applications from clothing to smart textiles.

2.3 Nanotubes

Nanotubes are hollow cylindrical tubes with their diameter in nanometers and length in microns (1D nanomaterial). Non-woven carbon nanotube (CNT) sheets are usually produced by the floating catalyst method, where thin webs of carbon nanotubes are made as sheets. Woven CNT fabrics are expensive and difficult to manufacture [4]. Carbon nanotubes are extensively used in textile and fashion industry due to their hollow nature providing high surface area, lightweight, excellent thermal and electrical conductivity aids in the integration of smart and electrical gadgets in clothing that increases its performance, functionality and safety. Development and exploration of different CNT materials and its incorporation in various applications can pave the way for potential advancement in invention of various smart and multi-functional textiles products.

2.4 Nanolayers

Nanolayers (2D nanomaterials) are very thin films or coatings with a nanoscale thickness. The inherent properties of the fabric are not affected by nanolayers when applied properly and carefully at controlled thickness. Bulk coating and uneven distribution of these materials over the fabric surface can stiffen the fabric, affecting the breathability and feel. The surface of the polyester can be modified by depositing polyelectrolyte nanolayers using a technique called layer-by-layer fabric deposition, where thin layers are deposited one layer over another in a controlled manner. This process smoothens the fabric surface and increases water retention [5]. A monolayer of graphene is coated over textile fibres by the vapour deposition method, this very thin single layer of graphene imparts transparency to fibres. Due to the presence of delocalised pi electrons in graphene, the graphene-coated fibres can conduct heat and electricity, acting as an electrode in a smart textile [6]. Selection of raw materials is one of the key aspects of sustainability in nanolayers. Sustainable production of nanolayers also focuses on lowering the energy utilised during synthesis and deposition. The usage of bio-based nanomaterials in films and coatings can reduce environmental harm and create a circular lifestyle for nano coatings.

2.5 Nanocomposites

Nanocomposites come under the category of 3-dimensional structures whose dimensions are more than 100 nm. Nanomaterials, nanotubes or nanofibres are embedded in the polymer matrix to form nanocomposites. The nanocomposites show numerous properties based on the filler nanomaterial, such as high performance, high strength, conductivity, antimicrobial resistance, UV resistance, etc. The properly engineered nanocomposites withstand frequent washing due to their increased durability. These composites are particularly designed for one or more properties which depend on the type of polymer matrix and nanomaterial used as a filler. To achieve the required end application, it is crucial to integrate a correct nano filler in the matrix. The usage of natural polymer like cellulose as a polymer matrix and eco-friendly nanofillers can be a sustainable approach which aligns with circular fashion goals.

3 Nanotechnology in Dyeing and Printing

Nanotechnology has brought various advancements in the textile dyeing and printing sector by optimising the performance of textile material, minimising environmental impact. The nanoparticles very effectively interact with fibres and fabric due to their incredibly small size and large surface area, allowing better dye absorption. Nanotechnology enables the development of superior inks with better penetration and durability. The key benefit of nanotechnology is sustainability, which ensures lower dye and water consumption and lessens the use of harmful chemicals. As the fashion industry moves towards a sustainable approach, the involvement of nanotechnology in the dyeing and printing process improves performance, and it also promotes sustainability and eco-friendly dyeing and printing processes.

3.1 Nanotechnology in Dyeing

Nanotechnology acts as an alternative to traditional dyeing, focusing on minimising waste energy consumption, chemical consumption, and promoting eco-friendliness. Nanoparticles exhibit multifunctional properties like antibacterial, UV resistance, self-cleaning, etc. When these particles are involved in the dyeing process, they not only improve dyeing efficiency and dye adherence to fabric but also can enhance the functionality, based on the type of nanomaterial used in the process. With the advent of nanotechnology, new solutions have emerged for dyeing fibres like polypropylene, which are very difficult to dye using conventional methods due to their crystalline nature, hydrophobicity and absence of reactive sites. The research carried out by Fan and Mani [7] suggests how nanotechnology has enabled the dyeing of polypropylene by dissolving nanoclay particles that have affinity to both ionic and non-ionic dyes in the polymer matrix. Even after the extrusion of fibres, the nanoclay particles retain their properties for the adsorption of dye. Another study on nanotechnology-based natural textile dyeing using curcuma (turmeric) for natural fabrics addresses the key limitation of conventional natural dyes. Compared to original dye particles, nano sized curcuma paticles show high K/S values (high colour strength) due to their high surface area ensuring better affinity to fabric especially in protein fibres like wool, silk and having high surface area, strong adherence and affinity to textile material potentially eliminate the usage of mordant like alum which makes the process simple and reducing chemical usage. Even at the very low concentrations, nano-sized curcuma shows better colouration than the high concentration of original dyes, leading to reduced dye usage [8]. Nanotechnology improves the dyeing process by converting curcuma into nanosized particles, offering a sustainable approach to natural dyeing.

The research conducted on the usage of nanotechnology for silk dyeing in 2011 suggests the usage of polyurethane acrylate binder along with TiO2 nanoparticles for increased colour strength (K/S value) of silk fibres. In this process, TiO2 nanoparticles act as a catalyst, enhancing the dyeing process by increasing the rate of binder polymerisation, resulting in increased uptake of pigments and improved colour [9]. In addition to this, a large amount of dye effluents, which include colouring and non-colouring matters released into the environment without pretreatment, causing serious environmental threats, can be prevented by using nano-sorbents to remove the organic and inorganic waste from aqueous medium by the process of adsorption. This process of removal is made even more sustainable with the help of biobased nanomaterials [10]. Nanotechnology also helps in dye decolourisation by using zinc oxide nanoparticles (ZnO). Under the UV light, these ZnO (zinc oxide) nanoparticles break down the synthetic dyes by generating OH groups. The study tested ZnO nanoparticles on 3 dyes, Congo red (88% decolouration), acid blue 9 (73% decolouration), and Coomassie Brilliant Blue R-250 (70% decolouration) [11]. The role of zinc oxide nanoparticles in decolouration has various advantages, like increased catalytic efficiency due to high surface area, cost-effectiveness and stability compared to catalysts like TiO_2 in specific conditions. According to [10], nanoparticles have been effectively utilised in the adsorption of dye pollutants from water medium in the last three decades. Nanotechnology contributes to both improved dye uptake and helps in promoting eco-friendly dye removal. Improved colour strength and fastness can be achieved by utilising nanoparticles with high surface energy, high surface area and reactivity. Additionally, nanotechnology provides sustainable methods for treating dye effluents from textile industries by breaking down harmful dye compounds into harmless by products. Nanotechnology not only improves efficiency in dyeing but also supports eco-friendly textile production, minimising pollution and conserving the quality of water.

3.2 *Nanotechnology in Printing*

Nanotechnology in textile printing has introduced new quality and precision levels to fabric. By using nanoparticles in pigments, the colour brightness, saturation, and durability of the print can be improved. Nanomaterials usage in printing also reduces the need for pretreatment and post-treatment in certain cases, making it an efficient solution in the modern production of textiles. Nanotechnology helps in various printing techniques like digital printing, pigment-based printing, and enhancing the properties of conventional pigments. Digital printing is a significant innovation in the textile sector because of its easy process and handling that does not require plate making like traditional textiles, and is employed for various fabrics, making the process faster and easier. Nanotechnology helps digital inkjet printing in several ways by using nanopigment inks. Nanopigments are on the nanoscale, extremely smaller than traditional pigments. Nanosized pigments allow improved durability, colour fastness, and enhanced image sharpness due to their small size. Additionally, nanoink does not clog the pores of the nozzles. Nanoinks are suitable for a wide range of fabrics and do not require extensive processing, eliminating the limitations of traditional printing [12]. Many innovations have been made to improve the functionality of the fabric through different printing methods by integrating it with nanotechnology. The involvement of nanotechnology in printing potentially eliminates the need for post-treatment and finishing of textile material to achieve certain properties, reducing the chemicals involved in the process. One such technique has been mentioned by Swelam [13]. The study specifically explores how silver nanoparticles can help to improve the functionality of the fabrics made from cellulosic fibres like cotton and linen when printed using natural dyes such as tea extracts. The high concentrations of silver nanoparticles showed deeper colour shades, self-cleaning properties, durability, UV resistance, and antimicrobial properties. Natural dyes are eco-friendly and reduce environmental toxicity, embedding nanoparticles like silver, adding effectiveness without the need for extra chemical treatments, and promoting sustainable textile production. The integration of CNT in the printing inks during inkjet printing offers good electrical conductivity that makes them suitable for printing sensors and wearable electronics in sportswear, health Monitoring, and fashion. This method is sustainable because it supports a solvent-free process at low temperature [14]. The fusion of nanotechnology and printing opens doors to many innovations and advancements and supports the vision of sustainable fashion by minimising waste and utilising eco-friendly nanomaterials.

4 Nanotechnology in Textile Finishing

Textile finishing is applied to fabrics after the knitting or weaving process to increase their texture, performance, and appearance. It is the final stage of the textile manufacturing process. This enhances the durability, functionality, and various physical and chemical properties of the fabric. The finishing process can be of a chemical or mechanical method:

- In chemical finishes, chemicals are used to enhance the desired properties of the fabric
- In mechanical finish, it uses physical processes to finish the fabric, which enhances the texture, durability, etc.

The use of nanotechnology in textile finishing has made the process much easier, effective, and made the process more sustainable. It also provides benefits in terms of eco-friendliness, energy and water saving, and the economy. Multifunctional finishes can also be achieved with the help of nanotechnology, where two or more different finishes can be given. Nanoparticles used and their application methods, and most importantly, the application of nanofinishing in UV protection, anti-microbial protection, and water repellent textiles are discussed in this topic.

4.1 Nanoparticles Used in Finishing

The finishing of textiles involves several types of nanoparticles. The most common are Zinc oxide, silver oxide, and titanium dioxide. Due to the economy, metal oxide nanoparticles are more preferred than nano silver. Chemically, zinc oxide and titanium dioxide are nontoxic and photo-catalytically oxidising agents stable at elevated temperatures. Fabrics containing nanoparticles are enhanced to accomplish diverse functional performance.

4.2 Methods Used for Nano Finishing

These are some of the common methods which is used for the application of nanoparticles in textile materials.

- Sol–gel method
- Dip coating method
- Electro-spraying process
- Layer-by-layer deposition

4.3 Application of Nanofinishing in Textile Materials

The application of nanofinishing in the textile industry helps to impart special properties that help in protecting and increasing the quality and performance of the textile material. Some of the most common types of nanofinishing used in the textile industry include antimicrobial finish, water repellent finish, UV protective finish, antistatic finish, wrinkle resistance finish, etc. Among these finishes, we are going to discuss the antimicrobial finish, water repellent finish, and UV protective finish in detail.

4.3.1 Antimicrobial Finish

Antimicrobial finish is applied on textile material to inhibit the growth of microorganisms such as bacteria, fungi, etc. It also prevents odour caused by bacteria. This enhances the hygiene, comfort, and durability of the textile, especially in medical and sports areas where bacterial growth is high and prevention of growth is highly needed. Antimicrobial finish is used in clothing, healthcare, and home textiles. Antimicrobial finishes can be applied to textile materials with the help of nanotechnology using nanoparticles such as silver nanoparticles (AgNPs), Titanium dioxide (TiO_2), Zinc oxide (ZnO), and several other particles. Metals like silver, copper have high antimicrobial properties, and the use of these materials can inhibit the growth of bacteria, fungi, and prevent odour. Hence, in nanotechnology, silver nanoparticles are used for the antimicrobial finishing in textiles. Among the various sectors, the Textile industry is the second most significant sector that uses nanosilver for antimicrobial properties [15]. Silver nanoparticles have strong antimicrobial properties that effectively reduce bacterial and fungal growth. Due to the larger surface area of nano silver particles, it increases the contact area with micro-organisms like bacteria and fungi, and thus it inhibits the growth of bacteria and fungi in the textile materials [16]. Likewise, another study used nano silver particles to give anti-microbial finishes on textile materials. The aqua extract of lemon leaves with a solution of silver nitrate was used, and their results suggest that the lemon leaf extracted silver nanoparticles employed to treat cotton and silk fabrics displayed durable antifungal finishes [17]. The applied nanoparticles, which form a coating on both cotton and silk fabrics, prevent fungi growth and increase the durability of silk and cotton fabric, which decays easily due to the growth of microorganisms. Titanium dioxide, the other nanoparticle, is used in antimicrobial finishing. A recent study done by Abdel et al. [18] used titanium dioxide nanoparticles to create antimicrobial polymer fabrics. It was applied with the help of polyacrylic acid binder at high temperature, and the samples were tested against microorganisms to assess antimicrobial performance, and its results suggested that the fabric treated with titanium dioxide has significant antimicrobial activity against several bacteria, yeast, and fungi. TiO_2 nanoparticles have such qualities that inhibit microbial growth, and when applied to the fabric, the nanoparticles become engaged with any microbes that access the fabric surface. It also increases durability, washing resistance, infection and longevity of the textile material. The use of nanotechnology in antimicrobial finishing has a significant effect on textile materials and has less environmental impact than other chemicals and thus imparting sustainability to the apparel industry.

4.3.2 UV Protective Finish

Ultraviolet (UV) protection is an important feature in modern textiles, especially for outdoor clothing and furnishings, as it helps shield users from harmful sunlight and prevents fabric damage. Application of nanotechnology for enhancing the properties of UV protection in textile material is carried out with inorganic nanoparticles like titanium dioxide (TiO_2) and zinc oxide (ZnO). Nanoparticles are effective as a blocking layer for UV due to their chemical stability, non-toxicity, and capability of absorbing and scattering ultraviolet radiation. The small size of the nanoparticles also increases the surface area to provide better protection, and according to Rayleigh's theory, nanoparticles of size between 20–40 nm will optimally scatter UV radiation (200–400 nm) [19]. The study by Kathirvelu et al. [20], where ZnO nanoparticles were treated on fabrics, observed that the surface of the fabric was dispersed with the ZnO and the absorption of UV rays took place, thereby protecting the human body from harmful rays. Treatments containing TiO_2 and ZnO, and indeed ZnO Nano rods, were shown to significantly increase the Ultraviolet Protection Factor (UPF) of polycotton and cotton fabrics with the effect remaining up to 50 washes [21]. Hence, the use of these nanoparticles, TiO_2 and ZnO, can help to protect human skin from harmful radiation and the diseases caused by UV rays. These nanoparticles can be applied to fabrics through various application methods without changing the fabric's feel or breathability. As this technology continues to improve, it is likely to replace conventional UV treatments in the production of sustainable textiles.

4.3.3 Water Repellent Finish

Water repellent finish is a treatment that reduces the penetration of water in the textile and hence makes the fabric resistant to water. In the fabric, the water rolls down the fabric instead of soaking in. These water repellent fabrics are inspired by the lotus leaf, where water does not stay on the leaf. Hydrophobicity means the materials that repel water. If the contact angle of water is above 90°, the material is said to be hydrophobic, while if it is above 150°, it is said to be superhydrophobic [22]. Hence, hydrophobicity should be imparted to make the fabric water repellent. Hydrophobicity can be obtained by increasing the surface area of the material. Nanotechnology can be used to make the textile materials water repellent without any environmental impact than conventional methods. Nanosilica particles are mainly used in this process. A study tells about the use of Nanosilica in the eri silk, the Nano silica is obtained from the rice husk, and applied to the fabric of Eri silk at different concentrations, and it is tested using different methods like contact angle, water absorbency test etc., and the results obtained indicate that there is an increase in the contact angle while there is an increase in the concentration of the nanosilica and the water absorption test, there was decrease in the time of absorption of ten water droplets in one minute at 1 percent nanosilica concentration [23]. Hence, the use of nanosilica has a greater impact on the water repellency. This also increases the durability, hydrophobicity, and breathability in the fabric and is an efficient method for a better sustainable future.

5 Nanotechnology in Smart Textiles and Wearable Technology

The textile materials that are manipulated to sense and respond to the external environmental conditions are called smart textiles [24]. According to Tao, there are three types of smart textiles classified based on the way they respond, namely passive smart textiles, active smart textiles, and very smart textiles. Innovations are being made in smart and wearable textiles, which is rapidly growing its popularity in health, fashion, military, and fitness. Nanomaterials like carbon nanotubes (CNT) and graphene act as a core element for the innovation of sensors and smart textiles. Nanotechnology plays a major role in improving the capabilities of the smart textiles without affecting their flexibility and appearance by working at the nanoscale, which is discussed below in detail.

5.1 Passive Smart Textiles

Passive smart textiles can sense the environmental changes [24]. For example: graphene enhanced silk fibres [25]. As reported by the researchers at Tsinghua University, feeding silkworms with graphene nanosheet-infused mulberry leaves produced silk fibres with improved mechanical properties and electrical conductivity. These fibres can sense environmental changes but do not respond to them actively [25]. Multifunctional finishes with nanomaterials come under the category of passive smart textiles. For instance, antimicrobial finishes coated with silver nanoparticles inhibit bacterial growth when they come in contact with the surface. They do not sense only momentarily, they are active all the time in the same way. There is no involvement of change in behaviour or decision making, so multifunctional finishes are considered as passive smart textiles. Passive smart textiles also include antimicrobial, water repellent, stain resistance, UV finishes, etc., all of which enhance the properties and performance of the fabric, improving the longevity, reducing the need for frequent washing, saving water and energy, and reducing frequent replacements. The passive smart textiles with multifunctional finishes often align with sustainability and eco-friendly fashion practices.

5.2 Active Smart Textiles

Active smart textiles can sense and respond to external stimuli [24]. Graphene-embedded smart textile is an example of it. In recent years, the development of adaptive smart textiles with the help of nanotechnology has opened doors to innovative applications in personal management and wearable technology. A recent study in this field suggests the use of graphene for developing climate-adaptive clothing. Multiple nanolayered graphene is coated over fabric like nylon, and then it is connected to a power source where a small amount of voltage is applied to change the amount of infrared light the graphene emits. Usually human body emits heat in the form of infrared radiation. The graphene has a natural ability (without voltage) to release heat. When a small voltage is applied, the normal interaction of graphene with heat changes. When the body heat rises due to outside temperature, the graphene-coated fabric releases infrared radiation outwards to keep the body cool. During winter seasons, with the supply of small voltage, graphene blocks the infrared radiation from escaping, retains heat, and keeps the body warm [26]. Just imagine a jacket that can keep the body cool in summer and warm during winter by escaping and retaining heat. Graphene coating on the fabric blocks or releases heat depending on electrical signals. The graphene-enabled adaptive textiles are at the research stage and have not been mass-produced yet. The researchers showed that these textiles are wearable, fast-acting, flexible, and do not require a large amount of voltage supply for their use. Integration of nanotechnology, like graphene, in adaptive textiles shows high potential for sustainably by reducing the requirement for heavy clothing or multiple layers to stay warm or cool. Garments can adjust according to temperature, so people could wear only one garment instead of many, reducing on usage of material and energy consumption. And graphene is a long-lasting material that can potentially reduce textile waste. Utilisation of these innovations can lower the reliance on solutions like indoor heating or cooling, which require heavy energy-making fashion smarter and environmentally friendly.

5.3 Very Active Smart Textiles

Very smart textiles can sense, respond, and adapt to external stimuli [24]. Electronic textiles are smart textiles integrated or embedded with sensors, antennas, LEDs, batteries, etc. that help in health monitoring, wireless communication via antenna, etc. Electronic textiles are considered very smart textiles because they can transmit, process, and store data, beyond just sensing, and nanotechnology assists in making these textiles flexible, wearable, and also lightweight (For eg, wearable graphene antennas). According to Ibanez et al. [27], the 2D nanomaterials, such as multilayered thin graphene coatings, are used to manufacture these wearable antennas by creating a multilayer of graphene with the aid of a nanofabrication process called chemical vapour deposition (CVD method). Then, after further processing, graphene is cut into a specific shape and laminated onto a cotton fabric over the ultrathin adhesive layer and laminated onto the textile. The SMA connector is fixed at one end of the antennas, allowing it to connect to external instruments. The advanced nanomaterials, like quantum dots, organic semiconductors, and electroluminescent particles, directly emit light when electrically stimulated. They are integrated through coating, screen printing, or embedding nano-thin light-emitting layers onto fabric. These materials sense electrical input, respond by emitting light, and adapt to body movements like bending, stretching, without losing light-emitting ability [28]. Nanotechnology-enabled wearable and flexible light-emitting textiles can be applied in fashion that combines functionality with aesthetic appeal. Nanotechnology is used in light-emitting textiles where the textile is made wearable, lightweight, and aligns with sustainable fashion goals by providing an alternative to traditional methods that involve the usage of lighting and electronic systems in textiles. Incorporation of nanofibres such as carbon and polyacrylonitrile or polyaniline nanofibres helps in producing Triboelectric nanogenerator textiles and gas sensing textiles, respectively [29]. Nanotechnology is revolutionising the textile and fashion industry by enabling fabrics to be smart and interactive without compromising comfort and originality. Nanotechnology helps the designers to blend aesthetics with technology. This fusion opens countless possibilities for sustainable and futuristic personalised fashion.

6 Biobased Nanomaterials and Their Role in the Fashion Industry

Biobased nanomaterials are derived from natural resources like animals, plants, and fungi. Some common biobased nanomaterials are jute nanofibres, nanocellulose, nanochitosan, and banana fibres. These materials are biodegradable, harmless, non-toxic, and possess special properties like mechanical strength and high surface area, making them suitable and valuable for various applications, especially in textiles.

6.1 *Nanochitosan*

Chitosan is extracted from a natural polymer called chitin found in shrimp shells and fungi, and these extracted polymers undergo various processes to manipulate them to the nanoscale. Nanochitosan and silver nanoparticles (optional) treated cotton fabrics exhibit long-lasting antimicrobial and antifungal properties, which are useful in medical wear and daily fashion wear [30]. Chitosan incorporated cotton fabrics have improved dye uptake for anionic dyes due to the cationic nature of chitosan [31]. It acts as a functional finish by enhancing breathability, UV protection, and has deodoriser properties, making it ideal for sportswear, healthcare textiles, and active wear.

6.2 *Nanocellulose*

Nanocellulose is derived from sources that are rich in cellulose, like cotton, wood pulp, and banana peels, and it undergoes acid hydrolysis to produce very thin, needle-like nanocellulose fibres with high strength and surface area. It supports sustainable fashion by acting as a dye carrier and improving dye fixation, eventually reducing the water usage during the dyeing process. Nanocellulose can be made to combine with polypyrrole, causing conductive textile modification, tailoring it for smart fashion textiles. It imparts various properties to the treated material such as better air permeability, absorbency, adhesion, stability, and combining allicin with nanocellulose provides durable antibacterial activity, making the fabrics suitable for variety of applications like hygiene clothing, medical garment, shape retention textiles, and everyday clothing wears [30].

6.3 Banana Nanofibres

Banana nanofibres are extracted from the banana peels or pseudostems. The extracted fibres undergo acid or alkaline treatment to eliminate noncellulosic parts and are further processed to create the fibres at the nanoscale. Nanofibres act as a reinforcing material in composites, improving fabric toughness and thermal stability. Banana nanofibres offer potential use in eco-friendly fashion composites and act as a sustainable alternative for reinforcement in textile composites supporting sustainable fashion [30].

6.4 Jute Nanofibres

Jute nanofibres are extracted from the jute plant and treated further to reduce them to the nanoscale. Stronger composites can be produced by blending polymers like PLA have high tensile strength, elasticity, and act as stronger composite materials. Jute nanofibres also absorb harmful chemicals from textile effluents due to their high surface area and natural absorbent properties, reducing the environmental footprint of garment manufacturing [30].

7 Nanotech Biomimicry in Sustainable Textiles

Nature is a supreme creator and offers enduring solutions through millions of years of evolution. From the self-cleaning lotus leaf to shark skin, the natural structures exhibit unique functionalities. Biomimicry is the study of these models and phenomena, then applying their concepts and principles to solve human problems. In the textile industry, the role of biomimetics can be made even more powerful by combining it with nanotechnology, allowing scientists to replicate the nanostructures present in nature on the fabric surfaces. This fusion has led to innovations like lotus leaf-inspired water repellent textiles, shark-mimicking drag-reducing fabrics, spider silk-inspired high-strength fibres, etc., contributing to the advancement of sustainable and intelligent clothing.

7.1 Lotus Leaf Effect: Super Hydrophobic Surfaces

Lotus leaves have nanostructured surfaces that are hydrophobic (repel water) and have self-cleaning properties. Nanoparticles like fluorocarbons, silica, and nanosilica-carbon hybridisation are used for mimicking these nanostructures, which makes the textile surface self-cleaning, stain-resistant, and water repellent. The incorporation of nanomaterials in biomimetics lowers the washing frequency and Detergent use, conserving water and energy [32].

7.2 Butterfly Wings Inspired Morphotex Fibre

Morphotex is a synthetic fibre that produces colour without using dyes. It is inspired by the vibrant colours on the wings of the morpho didius butterfly. Morphotex fibres use nanotechnology to mimic the microstructure of butterfly wings, creating colours without any kinds of dyes or pigments. These fibres produce vivid colours through structural design instead of pigments by layering the polymers at the nanoscale that reflect light in specific ways [33]. This eliminates the need for chemical dyeing, reduces water pollution, conserves energy, and supports sustainable fashion.

7.3 Shark Skin Effect

Shark skin has micro ridges that prevent drag and reduce microbial growth. E.g., Speedo Fast Skin is a high-performance swimwear that reduces the drag mimicking the skin of a shark and sharklet micro pattern with an antibacterial surface design inspired by shark skin [32]. These textiles use nanotechnology to replicate the tiny riblet structures found on shark skin. The study by Weerasinghe et al. [33] explores how nano manufacturing methods like nano weaving help in replicating these patterns directly onto the fabric without performing any traditional antimicrobial treatments. Modifying the surface, rather than using additional chemical treatments, reduces the chemical waste. This innovation supports safer, cleaner, and long-lasting garments.

7.4 Spider Silk and Nanofibre Spinning

Spider silk is a natural fibre spun by spiders with extraordinary strength, lightness, and elasticity. This natural model has inspired scientists to replicate its spinning process with nanotechnology, creating high-performance fibres using minimal energy through nanoscale spinnerets. This innovation supports sustainable fashion by offering strong materials that are strong and consume low energy [32].

8 High-Performance Textiles

Nanomaterials are minute particles that have a considerably large area-to-volume ratio, which makes these materials chemically reactive and effective. These help improve the durability, flexibility, repellency, and overall functionality of the fabrics, contributing to their high performance. Major use of nanoparticles can enhance the performance efficiency in fabrics, which can be seen in some of the fields like sports, military, eco textiles, etc.

8.1 Sports Textiles

Nanotechnology is rapidly changing the sport textiles with improved performance and comfort characteristics. The incorporation of nanomaterials and nano-finishing has resulted in better thermal control, moisture management, breathability, flexibility, odour resistance, and UV protection. Regarding thermal management, clothing developments incorporate nanofiber and nanocomposite structures which permit superior radiative cooling from our body meta fabrics created from titanium oxide-polylactic acid (TiO_2-PLA) or PTFE nanofibers work well and outperform typical fabrics by greater than 4° to 8 °C [34]. Additionally, in cold environments, nanomaterials such as ZnO nanoparticles achieve excellent thermal insulation through heat transfer, which makes garments suitable for extreme activities and environments. For moisture management, nanofiber composites and recently introduced Janus fabrics can efficiently wick away sweat from contact with skin and keep the athlete dry and comfortable. TiO_2 nanoparticles will maintain UV protection from sun exposure, while silver nanoparticles can be used for odour control and also provide antibacterial qualities. Importantly, after extended wear and washing, nanotechnologies will retain garments' properties of durability, breathability, and flexibility. The nanotechnology in improving sports performance and fabrics will emerge with a new wave of capacity and multi-functional fabrics that meet the current performance requirements of athletes to maintain comfort, keep athletes safe, and facilitate improved capabilities in a variety of sporting environments.

8.2 Military Textiles

Military textiles have moved quickly in recent years because of the increase in military spending globally and the need to give soldiers greater levels of protection when operating in extreme environments. Nanotechnology is enabling the development of next-generation military textiles that will be lightweight, provide better protection, be stealthy, multifunctional, self-healing, and environmentally sustainable. Materials like nanocellulose are being seen as promising for military applications, including packaging, energy devices, fire retardants, and filtration. A major innovation in military textiles is in EMI (electromagnetic interference) shielding for integrated electronics and communications. Innovative composites that use conductive fillers like graphene, carbon nanotubes (MWCNTs), black oxide magnetic nanoparticles, and nanofibers (e.g., aramid nanofiber, MXene/silver nanowire) are producing flexible, durable, and highly conductive fabrics with great thermal management capabilities [35]. Stealth technology is a second prominent area, where nanomaterials are facilitating new avenues for infrared thermal camouflage by controlling thermal emission or temperature. Approaches in this area include micro/nano-structured materials, quantum wells, electrochromic dyes, and phase change materials (PCMs). For instance, Kevlar nanofiber aerogel films with PCMs provide excellent thermal insulation properties, while also giving infrared stealth [36]. Similarly, the use of nanofibers and nanocomposites in protective clothing for ballistic, fire, chemical, and extreme weather protection is enabling products to be lightweight, as they are also better protective solutions. The use of nanotechnology has had a significant effect on the military by incorporating various functions in a single garment, which is multifunctional and effective, and these textiles have made a huge impact on the military sector.

8.3 Eco Textiles

These are an innovation in the textile industry to address the industry's considerable sustainability issues, such as high unavoidable carbon emissions, toxic waste, and resource drain. The traditional textile manufacturing process threatens to pollute the air and water while causing terrible soil erosion and adding greenhouse gases to the atmosphere. The industry is embracing eco or environmentally compliant materials as well as nanotechnology. Nanotechnology makes it possible to manufacture eco-friendly nanocomposite fibres using nanocarbon fillers, such as carbon nanotubes or carbon black, combined with conventional polymers (such as polyamide, polyethylene, and polyester). Sustainable nanocomposite fibres have applications in various fields, such as electronic textiles, antimicrobial textiles, military textiles, and self-healing materials, with lesser dependence on fossil fuels compared to conventional textiles. Another remarkable finding is the triboelectric nanogenerator (TENG), capable of collecting mechanical energy (e.g., motion) to generate electrical energy [37]. Porous nanocomposite fabrics filled with nano-Al_2O_3 provide an efficient, lightweight, and eco-friendly energy choice to power sensors and microelectronics embedded in textiles, respectively. Also, the regenerative and bacterial cellulose are new, fast-tracked materials. Bacterial cellulose, a biodegradable nanofibrous polymer, can be designed for specific applications in textiles, creating an eco-friendly alternative for the apparel and footwear industry. Hence, the use of eco-textiles with the help of nanotechnology will lead to an environmentally sustainable industry.

9 Environmental Impact

Nanotechnology offers vast opportunities in terms of textile functionality and performance, but it is also critical to comprehend the environmental cost and sustainability of using it. There has been a concern about the environmental effects of making and discarding nanomaterials and the possible release of particles during the life cycle of nanotechnology-enabled textiles. The case study by Voelker et al. [38] suggests that nanosilver release from fabrics can be environmentally safe if those particles are incorporated in the fibre, as surface coated fabrics have higher release of nanoparticles, which can be accumulated in water and soil. To address these issues, work on green synthesis methods to minimise the application of toxic chemicals and on the production of the desired material in an environmentally safe way by minimising energy and generating waste is ongoing. Studies on the environmental fate and toxicity of nanomaterials are also being carried out, and protocols for environmentally safe disposal of waste are being established. We can preserve the environment using introducing sustainable materials, e.g., biodegradable polymers, renewable resource-based materials, or recycled materials. Additionally, the textiles industry is now turning attention to a life cycle perspective and circular economy adoption that encourages and curtails the dangers of nanotechnology as an application enhancement to textiles that also enhances their environmental sustainability Another study examined the release of silver particles from textile fabrics using conventional and nano silver during washing, and the results show that the conventional silver textiles generally released more silver particles than nano silver textiles [39]. For instance, numerous coating technologies employ nanomaterials such as silica, clay, or cellulose-based materials to develop coatings that are biodegradable, nontoxic, and non-harmful.

10 Cost and Scalability

Nanotechnology has been a part of sustainable fashion by providing advanced functionalities in textiles like different finishes, such as UV resistance, antimicrobial, and water resistance finishes, as well as various smart textiles, high-performance textiles, and more, without compromising fabric comfort. A research study shows that even after 20 washes, the AgNP-treated cotton fabric exhibits sustained antimicrobial activity, reducing the need for frequent washing [40]. These enhancements can prolong the life cycle of garments and reduce the frequency of washing, thus contributing to environmental sustainability. Considering the textile industry, where there is a large use of nanotechnology and the inclusion of the technology in textile manufacturing processes, there is a need to reflect on what scalability and cost-effectiveness imply for nanotechnology. The expense of nanomaterials such as carbon nanotubes, titanium dioxide, and silver nanoparticles, and the requirement for specialised production methods, are the primary obstacles to cost-effectiveness and scalability. Some of the nanomaterial challenges have been to manufacture them in sufficient quantities to supply market demand with good quality, and to compete with the price of other fabrics. To this end, researchers and industry experts are working on the development of cost-efficient and scalable methods of production. Although the initial investment into nanotechnology-based textile production is larger today than traditional textile development, developing production techniques, as well as growing demand for textile products that have been enhanced with nanotechnology-type properties, can certainly lower the cost of production in the future. A study by Haque and Naebe [41] shows how plasma alters wool fibre surfaces at the nanoscale, allowing for quick, waterless natural dyeing (e.g., onion skin extract) that takes only one-third the time and energy of traditional procedures while boosting colour strength. Nanotechnology can lower the environmental impact of the fashion industry, thus, the future of nanotechnology and sustainable fashion is promising, with a demand for investment in safer, more cost, efficient and environmentally responsible methods for production.

11 Future Scope and Directions

Nanotextiles can be treated with molecular dyes and analyte-sensitive reagents. Considering microfluidics, it can be incorporated into thread-based conduits, which could be used for point of care diagnostic applications [42]. Shortly, various functional elements are ready to be incorporated seamlessly into textile structures. The application of nanotechnology in the textile industry has the potential to radically alter the technical textile sector. However, commercialisation of items based on nanotechnology involves several obstacles, caution is recommended [43]. Materials such as polymers and textiles must be developed to attain industrial applications. The major challenge is to ensure the dispersion and stability of nanoparticles within polymer matrices. Nanoparticles tend to agglomerate because of their exceptionally large surface area, leading to their effective incorporation into coatings, finishes, and nanocomposite developments. Desired nanoscopic characteristics can be integrated without any loss with the help of surface engineering of nanoparticles and their blending with functional surface-active polymers [44]. Nanotechnology and nanomaterials have provided intriguing properties for the textile sector. Their capacity to modify material at the nano level opens numerous possibilities to improve textile functionality and performance. Remarkable performance has already been made, resulting in high-performance fabrics with improved strength, durability, and water repellent property, along with unique qualities like antimicrobial activity, ballistic protection, UV protection, Self-cleaning, and electrical function. Despite these advances, significant challenges remain which are still need to be overcome, requiring concentrated efforts of industrialists, scientists, and the government to address the dispersion, environmental safety, and regulation regarding the use of nanoparticles. The future of research in this sector promises hope with continued breakthroughs in smart textiles that can detect and react to various environmental conditions, energy harvesting fabrics, and wearable devices embedded with nano sensors. Furthermore, continued research needs to be carried out to ensure sustainable and ethical application of nanotechnology in the textile industry.

12 Conclusion

The use of nanotechnology has shaped both the textile and fashion industry in different ways, which promotes sustainability. This technology has enhanced the properties of fabrics by improving their durability and functionality without altering their comfort. This chapter discusses the various nanomaterials used in textiles, such as nanoparticles, nanotubes, nanofibres, and nanocomposites. Nanotechnology has been used in the textile industry, starting from fibres, dyeing, printing, to finishing. This section highlights the application of nanotechnology-based approaches in the textile sector for its efficiency, performance, and sustainability. This also includes the integration of smart textiles in the form of active, or very active, smart textiles in the fabric, which can be used to detect and manage temperature, climatic adaptive clothing, and in health monitoring aspects. Biobased nanomaterials are used because it is easily biodegradable and non-pollutant to the environment, and are even used in bio-mimicking processes. Overall, this chapter concludes with the present usage of nanoparticles and nanotechnology in the textile industry for high performance and sustainability, and also discusses the opportunities for future research and development in this field.

References

1. Grand View Research (2023) Nanotechnology market size, share & trends analysis report. [Online], https://www.grandviewresearch.com/industry-analysis/nanotechnology-market-report. Accessed 23rd May 2025
2. Pokropivny V, Lohmus R, Hussainova I, Pokropivny A, Vlassov S (2007) Introduction to nanomaterials and nanotechnology. University of Tartu, Estonia
3. Ko FK, Yang H (2008) Functional nanofibre: enabling material for the next generations smart textiles. J Fiber Bioeng Inform 1(2):81–92
4. Chitranshi M, Chauhan D, Kubley A, Pujari A, Xu C, Chen D, Chaudhary S, Hou G, Bell G, Brandewie B, Kaneria R (2019) Pioneering carbon nanotube textile engineering & fashion technology. J Text Eng Fash Technol 5:89–92
5. Jantas R, Polowinski S (2007) Modifying polyester fabric surface with polyelectrolyte nanolayers using the layer-by-layer deposition technique. Fibr Tex Eastern Europe 15(2):97
6. Neves AIS, Bointon TH, Melo LV, Russo S, De Schrijver I, Craciun MF, Alves H (2015) Transparent conductive graphene textile fibers. Sci Rep 5(1):9866
7. Fan Q, Mani G (2007) Dyeable polypropylene via nanotechnology. In: Nanofibers and nanotechnology in textiles. Woodhead Publishing, pp 320–350
8. Ragheb AA, Tawfik S, Abd-El Thalouth JI, Mosaad MM (2017) Development of printing natural fabrics with curcuma natural dye via nanotechnology. Int J Pharm Sci Res 8(2):611–620

9. El-Molla MM, El-Khatib EM, El-Gammal MS, Abdel-Fattah SH (2011) Nanotechnology to improve coloration and antimicrobial properties of silk fabrics. Indian J Fibre Text Res 36(3):266
10. Osagie C, Othmani A, Ghosh S, Malloum A, Esfahani ZK, Ahmadi S (2021) Dyes adsorption from aqueous media through the nanotechnology: a review. J Market Res 14:2195–2218
11. Divya B, Karthikeyan CH, Rajasimman M (2018) Chemical synthesis of zinc oxide nanoparticles and its application of dye decolourization. Int J Nanosci Nanotechnol 14(4):267–275
12. Lin CH, Lin CM (2013) The study in the application of nanotechnology digital printing in clothing pattern design. Key Eng Mater 562:674–679
13. Swelam ASAA (2018) Usage of nanotechnology and natural dyes in the printing of cellulosic clothing fabrics. Int Design J 8(4):259–280
14. Eshkalak SK, Chinnappan A, Jayathilaka WADM, Khatibzadeh M, Kowsari E, Ramakrishna S (2017) A review on inkjet printing of CNT composites for smart applications. Appl Mater Today 9:372–386
15. Damuluri R, Babel R (2023) Nanotechnology for antimicrobial textile finishing–a review. Environments 7(8)
16. Wong YWH, Yuen CWM, Leung MYS, Ku SKA, Lam HLI (2006) Selected applications of nanotechnology in textiles. AUTEX Res J 6(1):1–8
17. Vankar PS, Shukla D (2012) Biosynthesis of silver nanoparticles using lemon leaves extract and its application for antimicrobial finish on fabric. Appl Nanosci 2(2):163–168
18. Abdel Salam KA, Ibrahim NA, Maamoun D, Abdel Salam SH, Fathallah AI, Abdelrahman MS, Mashaly H, Hassabo AG, Khattab TA (2023) Anti-microbial finishing of polyamide fabric using titanium dioxide nanoparticles. J Textil Colorat Polym Sci 20(2):171–174
19. Ghosh J, Repon MR, Anas NH, Noor T, Mia MN, Khan F, Sultana S, Rupanty NS (2024) Development of UV protective and antimicrobial CVC fabric using colloidal zinc oxide solution with carrot and orange peel extract. SPE Polym 5(4):576–589
20. Kathirvelu S, D'souza L, Dhurai B (2009) UV protection finishing of textiles using ZnO nanoparticles. Indian J Fibre Text Res 34(3):267–273
21. Gokarneshan N, Gopalakrishnan PP, Jeyanthi B (2012) Influence of nano finishes on the functional properties of textile materials. Int J Basic Appl Chem Sci 2(2):16
22. Zsidai L, De Baets P, Samyn P, Kalacska G, Van Peteghem AP, Van Parys F (2002) The tribological behaviour of engineering plastics during sliding friction investigated with small-scale specimens. Wear 253(5–6):673–688
23. Borah MP, Jose S, Kalita BB, Shakyawar DB, Pandit P (2020) Water repellent finishing on eri silk fabric using nano silica. J Tex Inst 111(5):701–708
24. Tao X (2001) Smart fibres, fabrics and clothing: fundamentals and applications. Elsevier
25. Wang Q, Wang C, Zhang M, Jian M, Zhang Y (2016) Feeding single-walled carbon nanotubes or graphene to silkworms for reinforced silk fibers. Nano Lett 16(10):6695–6700
26. Ergoktas MS, Bakan G, Steiner P, Bartlam C, Malevich Y, Ozden-Yenigun E, He G, Karim N, Cataldi P, Bissett MA, Kinloch IA (2020) Graphene-enabled adaptive infrared textiles. Nano Lett 20(7):5346–5352
27. Ibanez-Labiano I, Ergoktas MS, Kocabas C, Toomey A, Alomainy A, Ozden-Yenigun E (2020) Graphene-based soft wearable antennas. Appl Mater Today 20:100727
28. Cinquino M, Prontera CT, Pugliese M, Giannuzzi R, Taurino D, Gigli G, Maiorano V (2021) Light-emitting textiles: device architectures, working principles, and applications. Micromachines 12(6):652
29. Salman AA, Metwally FA, El-bisi MK, Emara G (2020) Applications of nanotechnology and advancements in smart wearable textiles: an overview. Egypt J Chem 63(6):2177–2184
30. Barik B, Maji B, Sarkar D, Mishra AK, Dash P (2022) Cellulose-based nanomaterials for textile applications. In Bio-based nanomaterials. Elsevier, pp 1–19
31. Zhou BCE, Kan CW, Sun C, Du J, Xu C (2019) A review of chitosan textile applications. AATCC J Res 6(1_suppl):8–14
32. Das S, Bhowmick M, Chattopadhyay SK, Basak S (2015) Application of biomimicry in textiles. Curr Sci 893–901

33. Weerasinghe DU, Perera S, Dissanayake DGK (2019) Application of biomimicry for sustainable functionalization of textiles: review of current status and prospectus. Text Res J 89(19–20):4282–4294
34. Syduzzaman M, Hassan A, Anik HR, Akter M, Islam MR (2023) Nanotechnology for high-performance textiles: a promising frontier for innovation. ChemNanoMat 9(9):e202300205
35. Ma Z, Kang S, Ma J, Shao L, Zhang Y, Liu C, Wei A, Xiang X, Wei L, Gu J (2020) Ultraflexible and mechanically strong double-layered aramid nanofiber–Ti3C2T x mxene/silver nanowire nanocomposite papers for high-performance electromagnetic interference shielding. ACS Nano 14(7):8368–8382
36. Xiao L, Ma H, Liu J, Zhao W, Jia Y, Zhao Q, Liu K, Wu Y, Wei Y, Fan S, Jiang K (2015) Fast adaptive thermal camouflage based on flexible VO2/graphene/CNT thin films. Nano Lett 15(12):8365–8370
37. Wang ZL (2017) On Maxwell's displacement current for energy and sensors: the origin of nanogenerators. Mater Today 20(2):74–82
38. Voelker D, Schlich K, Hohndorf L, Koch W, Kuehnen U, Polleichtner C, Kussatz C, Hund-Rinke K (2015) Approach on environmental risk assessment of nanosilver released from textiles. Environ Res 140:661–672
39. Mitrano DM, Rimmele E, Wichser A, Erni R, Height M, Nowack B (2014) Presence of nanoparticles in wash water from conventional silver and nano-silver textiles. ACS Nano 8(7):7208–7219
40. Jain A, Kongkham B, Puttaswamy H, Butola BS, Malik HK, Malik A (2022) Development of wash-durable antimicrobial cotton fabrics by in situ green synthesis of silver nanoparticles and investigation of their antimicrobial efficacy against drug-resistant bacteria. Antibiotics 11(7):864
41. Haque ANMA, Naebe M (2023) Zero-water discharge and rapid natural dyeing of wool by plasma-assisted spray-dyeing. J Clean Prod 402:136807
42. Yetisen AK, Qu H, Manbachi A, Butt H, Dokmeci MR, Hinestroza JP, Skorobogatiy M, Khademhosseini A, Yun SH (2016) Nanotechnology in textiles. ACS nano 10(3):3042–3068
43. Asif AKMAH, Hasan MZ (2018) Application of nanotechnology in modern textiles: a review. Int J Curr Eng Technol 8(2):227–231
44. Joshi M, Bhattacharyya A (2011) Nanotechnology–a new route to high-performance functional textiles. Text Prog 43(3):155–233

Advances in Sustainable Fashion: Zero Waste Design—A Multi-dimensional Approach

R. Surjit, P. B. Nehaasri, S. Aswathi, and M. Kavya

Abstract The fashion industry has a deep impact on our environment, with textile waste being a significant threat. We can overcome this issue by moving towards sustainable fashion. Sustainable fashion is the concept of developing a garment that causes minimal damage to our environment. Sustainable fashion uses eco-friendly practices. One of those is the zero waste design. The zero waste design is a tool as well as a process. It is a tool when techniques like 5Rs are used, and it is a process when designers and manufacturers rethink the processes and materials to create products that are durable and can either be recycled or reused. This chapter discusses how zero waste design can be used for promoting sustainable fashion. The concept of utilising zero waste design in designing and pattern making, production, disposal, and circularity is discussed in detail in this chapter. The chapter also highlights the advancements of the zero waste design in sustainable fashion and examines its potential for minimal wastage and the promotion of circularity. This chapter also discusses the technological innovations to implement the zero-wastage program, like the usage of advanced CAD to produce intricate patterns with optimal fabric usage. This chapter also emphasises the importance of material selection in achieving zero waste. The zero waste design allows to take a huge turn from 'take, make, dispose' to 'design, reuse and restore', thereby leading to sustainable fashion.

Keywords Sustainability · Zero waste · CAD · Recycling · 3D printing

1 Introduction

In this world, the fashion industry is burdened by environmental impacts like climate change and social crises, which have increased widely. The effect prompts us to adopt the concept of zero waste. Zero Waste talks about waste management strategy, which provides sustainability in fashion. The principle of this concept is to eliminate

R. Surjit (✉) · P. B. Nehaasri · S. Aswathi · M. Kavya
Department of Fashion Technology, PSG College of Technology, Coimbatore, Tamil Nadu, India
e-mail: rst.fashion@psgtech.ac.in

S. S. Muthu (ed.), *Sustainable Solutions in Textiles and Fashion*, Sustainable Textiles: Production, Processing, Manufacturing & Chemistry,
https://doi.org/10.1007/978-3-032-10925-5_5

waste, conserve resources, and create sustainable systems. The zero waste concept encourages ideas in designing, pattern making, production, disposal, and circularity. In pattern making, zero waste is widely accepted because the pattern-making process produces the highest wastage compared to all other processes. According to facts, more than 400 billion square meters of fabric is produced in a year for clothing, where 15 per cent of the fabric ends up as waste [1]. A technique is introduced in pattern making by applying mathematical equations to make patterns to save fabric wastage [2] and creating jigsaw-like arrangements in pattern making supports zero waste. According to the data, the fashion industry is the third most polluting industry in the world. To prevent the effect, the circular economy is introduced as one of the solutions, which includes reduce, reuse, and recycle. This can be used for minimising the wastage and providing a proper natural cycle. Disposal is another concept where zero waste can be achieved. According to facts, a large portion ends up in landfills or is incinerated, a very small percentage is reused or recycled, and the maximum level is discarded. To stop this, we should provide proper disposal, which is every human's responsibility. Designing is another area where the zero waste concepts combine in the fashion industry. After designing a fashion product, it should be draped to avoid fabric wastage, and following the jigsaw concept of pattern arrangements provides a zero-waste design concept. The production side of the zero-waste concept encourages using CAD software systems for pattern layout and cutting. In the dyeing process, the idea of natural dyeing techniques is introduced. All the zero waste concepts should be educated to the consumers, and also offering thoughts to brands who are interested in a sustainable fashion journey. The scope of using zero waste design in promoting sustainable fashion, utilisation of zero waste design in various apparel processes, and advancements in zero waste design are discussed in this chapter. The chapter also provides details on technological innovations aiding in achieving zero waste design and materials that can be used for achieving zero waste.

2 Zero Waste in Design

Zero waste in design refers to the thoughtful integration of sustainability principles into the fashion design process to minimise or eliminate textile waste. It is said that five criteria (aesthetics, fit, cost, sustainability, and manufacturability) must be balanced depending upon the design, context, and stage of development [3]. A zero waste approach should neither compromise garment appeal and comfort nor lead to unnecessarily high production costs due to overly complex constructions. While the primary focus may be on reducing fabric wastage and achieving zero wastage, design is a part of a broader sustainability framework. The fashion industry is inherently complex, and achieving sustainability requires flexible, multi-scale solutions. Therefore, zero strategies must move beyond isolated techniques and instead align with sustainable practices to create meaning for an inclusive experience in fashion from the designers to the end users.

2.1 *Digital Design and Virtual Sampling*

The fashion industry contributes between 5 and 10% of global greenhouse gas emissions, surpassing the emissions from international flights and maritime shipping combined. This is largely due to energy-intensive production, processes, and transportation of samples and garments [4]. One of the main causes of this problem is too much shipping between the buyer and the manufacturer, which can be avoided or reduced by leveraging advanced digital 3-D technology, such as CLO 3D, Optitex, where designers can visualise and modify garments in a virtual environment before any fabric is cut. This reduces the dependency on physical samples, minimizes textile waste, shortens the development cycles, and limits carbon emissions associated with production and logistics. Such technologies enable precision in fit and form, enhance collaborative workflows, and support the industry's broad move towards zero waste practices and eco-conscious design.

2.1.1 Role of 3-D Software in Sustainable Fashion

The incorporation of 3D software into fashion design and product development has emerged as a transformative step toward sustainability in the apparel industry. Sampling physically during the traditional fashion design process is extremely inefficient as it requires multiple iterations, which leads to excessive use of materials, resources, and increased carbon footprint. On the other hand, 3D virtual prototyping tools allow designers to visualize garments and make modifications in a digital environment, which mitigates the need for sewing or cutting fabric, working as an environmentally friendly option.

Advanced simulations for 3D design software such as CLO 3D, Optitex, and Browzwear allow for the exact representation of garment fit, drape, and construction, which improves accuracy while reducing dependence on physical samples. These advancements greatly decrease waste in textiles, cut carbon emissions from sample shipments, and enhance business agility. Improvements in interdepartmental communications, like marketing, production, and design, alongside the accelerated decision-making processes and streamlined lead times maximise supply chain efficiency.

Additionally, these 3D design tools shift business models towards sustainability by enabling functions like small-batch production, pre-orders, and made-to-measure, which minimise overproduction and unsold inventory. Although there are concerns regarding the initial adoption of these tools due to their technological constraints, interfaces and education programs tailored for digital learning are rapidly evolving to help resolve these issues.

Ultimately, 3D software plays a pivotal role in bridging creativity with responsible production. By enabling a shift from material-intensive processes to digital-first strategies, it offers the fashion industry a practical path toward zero-waste goals and long-term environmental stewardship [5].

3 Zero Waste in Pattern Making

Zero waste pattern making is a sustainable fashion design technique that aims to eliminate textile waste by strategically arranging pattern pieces to use the entire fabric without leaving scraps. Pattern pieces are arranged like puzzle parts, utilizing negative space efficiently to prevent any fabric from being discarded. Every inch of fabric, including small scraps, is incorporated into the garment either functionally or decoratively. Although they can be used for various purposes, they include some challenges and some of them are: require high planning accuracy and unconventional cutting methods, difficult to adapt zero waste patterns across multiple sizes without redesigning layouts, and may increase labour and construction time [6].

3.1 Pattern Tetris: Maximizing Fabric Yield

Just like the game Tetris, where pieces must fit together perfectly without gaps, zero waste pattern making relies on arranging garment pattern pieces so that they fill up the entire fabric width and length, leaving little to no unused space.

Example: Instead of cutting a sleeve, bodice, and pant leg separately, a designer may manipulate their shapes by covering the pattern of repeated shapes that fit together closely without gaps or overlapping.

3.1.1 Methods Used in Pattern Tetris

In zero waste pattern making, several innovative methods are employed to eliminate fabric waste while maintaining design integrity. One of the main methods is geometric pattern planning, which involves using simple shapes like rectangles and trapezoids. These geometric patterns are easier to align and stick together efficiently on the fabric, minimising unused space. Another method is negative space engineering, where designers and pattern makers carefully analyse the gaps between pattern pieces. Rather than leaving these spaces unoccupied, they are creatively repurposed into functional, decorative, or structural elements such as godets, gussets, or ties, thus contributing to the garment's design while also using up the excess fabric. Lastly, pattern integration plays a vital role by combining smaller elements into larger ones. This not only decreases the total number of individual components but also enables more flexible and compact pattern layouts, optimizing every bit of fabric. A practical example is joining a pocket with a side panel or dart seam in a jacket or dress [7].

3.1.2 Benefits and Challenges

Zero waste pattern making doesn't just help reduce fabric waste, but it also brings many other benefits. It inspires designers to think more creatively, which often leads to unique and interesting garment shapes, like folds or asymmetrical designs. It is good for the environment because it cuts down on fabric waste, which is one of the major pollution problems in fashion. This method keeps up sustainable practices and is great for the slow fashion movement, where clothes are made carefully, ethically, and often by local craftspersons. Most people today prefer buying clothes that have a positive story behind them, and zero waste design supports that idea. However, there are some challenges and disadvantages. It's challenging to make the same design in different sizes and designs without changing the layout, so designers might need to create new patterns for each size. Also, some fabrics, especially ones with prints or patterns, can be hard to use in this way. Therefore, labourers at factories might need extra training to make these garments because the way they're made can be different from usual [8].

3.2 Integrated Pattern-Design System

Integrated pattern design systems is a study of approach in fashion where the processes of designing, pattern making, and garment construction are combined into one. Instead of handling each stage separately, which is common, this method consists of all these techniques in one process. The result of the process is a more thoughtful design process that not only enhances creativity but also reduces fabric waste and supports zero waste processes. This system plays an important role in zero waste, where how a garment is designed directly affects how efficiently fabric is used. The pattern includes 4 information: namely, basic pattern information, description of the design, description of the final design, and performance of the final design [9].

3.2.1 Role of Integrated Pattern Systems

In a study related to pattern systems, designers begin thinking about pattern shapes, fabric width, and garment construction during the initial design phase, which leads to more innovative outcomes and less material waste. On account of this, patterns aren't just a technical step, they are part of the creative process. For example, a sleeve, collar, or even a pocket is not added later but is designed from the beginning to fit perfectly into the available fabric space in the patterns. This kind of thinking is necessary for sustainable fashion, in which minimizing waste is a major goal. Instead of drafting patterns in a traditional flat format, this technique removes sections of fabric to create new components directly in the garment. This experimental method does not use standard pattern development but allows the garment to take form organically, blending construction and design into one. On account of this, integrated systems can unlock creativity and move away from rigid, standardized clothing design.

By reducing the number of steps and materials needed, these techniques promote efficiency and sustainability and cut down on waste. This also means less energy is used in production, and garments often have a longer lifespan due to more thoughtful and creative design. Integrated systems line up with the values of the slow fashion movement, which encourages quality, longevity, and sustainability in how clothing is made.

Integrated pattern design systems are about merging creativity with sustainability. Through the study and research of this technique, we see how fashion can evolve into a more thoughtful and responsible industry. By designing with the end in mind and considering the fabric, construction designers can create garments that are both beautiful and sustainable.

3.3 *Subtractive and Addictive Pattern Logic*

Subtractive and additive pattern logic explores two different approaches to garment construction and pattern development in fashion design. Understanding these logics helps designers experiment with different forms, structures, and sustainability in unique ways. While traditional fashion design has mostly based on additive pattern logic, modern practices and particularly in zero waste fashion subtractive logic is introduced as a powerful alternative.

3.3.1 Additive Pattern Logic

Additive pattern logic is the most commonly used method in traditional pattern making. In this approach, designers start with a flat, mostly rectangular piece of fabric and place pattern pieces on top, like sleeves, collars, waistbands, and other components, and then cut them out and sew them together. The focus is on the construction of the garment from individual smaller pieces to form the whole. This method is organised and easy to produce, making it ideal for mass production. However, one major disadvantage is the high amount of fabric waste that is often left over, as excess material is generated. Additive logic often overlooks sustainability because the fabric layout is not optimized for waste reduction. While it allows for detailed control over shape and fit, it is often not the most environmentally efficient [3].

3.3.2 Subtractive Pattern Logic

Subtractive pattern logic starts with a whole fabric and then cuts sections to create shape, volume, or openings. Instead of building the garment from separate pieces, the garment takes shape through strategic subtraction. This is a non-traditional way of approaching the end pattern pieces. The result is often a garment with unique drapeability, layers, and omission of unwanted spaces that challenge conventional fashion aesthetics. Subtractive logic is closely related to zero waste practices because it aims to use the fabric as a complete whole and design within its limits, rather than cutting pieces that are left behind as waste.

This kind of pattern-making encourages designers to embrace the unpredictable, allowing patterns to emerge through experimentation rather than strict planning. This makes subtractive logic more innate and less rigid and is ideal for designers looking for creative freedom and sustainable outcomes. In comparison, additive logic, which is structured and practical, often prioritises fit and function over sustainability unless carefully managed.

To sum up, both logics offer unique benefits. In additive logic, it is ideal for accuracy and scalability, while subtractive logic is valuable for innovation and waste reduction. The challenge and opportunity for modern designers lie in understanding when and how to apply each method or combine them for both aesthetic and sustainable success [10].

3.4 *Digital Tools for Waste-Free Layouts*

The increase of digital technology has notably transformed design and manufacturing processes, particularly in the pursuit of sustainability. Waste-free layouts, also known as zero-waste design, aim to eliminate material waste at the starting stage, reducing environmental impact and enhancing efficiency. Numerous digital tools have emerged to support this goal, allowing designers and manufacturers to plan, visualize, and finalize layouts with minimal material loss. One of the key advancements in this technology field is the use of Computer-Aided Design (CAD) systems. CAD programs like CLO 3D and Gerber AccuMark allow precise pattern drafting and modification, which is important in minimising offcuts and optimising fabric usage. Digital tools enable a deeper integration of design and production, enabling zero-waste principles to be applied from the initial planning phase. Another major advantage is the use of nesting software, such as Optitex and Lectra, which automates the layout of pattern pieces to ensure the most efficient use of material. These tools depend on complex algorithms that consider pattern shape, fabric grain, and other constraints to reduce waste [11]. Digital simulation is emphasised as an essential tool for visualizing how design choices impact waste and sustainability. Simulation tools also play a crucial role in virtual prototyping, which reduces the need for physical samples abruptly and decreases material usage during the development stage. Moreover, 3D visualization platforms like Browzwear and Marvelous Designer allow designers to assess fit, drape, and aesthetics digitally, facilitating more informed decisions without producing multiple physical prototypes. Integration with PLM (Product Lifecycle Management) systems further aids in tracking material usage, enabling a holistic view of the production cycle, which supports sustainable practices. Overall, digital tools not only enhance creativity and precision in layout planning but also serve as a turning point for sustainability and zero-waste practices in modern design industries.

3.5 *Pattern Sculpting for Seamless Assemblies*

Pattern sculpting for seamless assemblies is a crucial concept in fashion design, focusing on creating garment patterns that integrate smoothly without visible seams or stitches. This technique ensures both aesthetic appeal and functional uses in clothing construction [8].

Important methodologies include dart manipulation, adding fullness, and contouring. Dart manipulation allows the fabric to fit body curves, adding fullness gives volume for design elements like gathers, and contouring shapes the garment to the body's shape [12].

a. **Dart Manipulation**

Darts are used to shape a flat piece of fabric into a three-dimensional form that fits the body's curves. Darts are especially useful in the bust, waist, and hips. Dart manipulation involves rearranging, combining, or transforming darts without affecting the fit. The techniques of dart manipulation include:

- Slash and spread: Cutting and spreading the pattern to shift the dart to another location (e.g., from the side seam to the neckline).
- Pivot method: Rotating the dart around the bust point to move it without altering fabric dimensions.
- Converting darts to design features, Such as pleats, gathers, or style lines, to create visual interest while maintaining fit.

b. **Added Fullness**

This technique involves adding extra volume to a garment for design or movement. It's essential for:

- Gathers and ruffles: Often added to sleeves, bodices, or skirts.
- Flared hems or bell sleeves: Created by spreading the pattern outward [13].

4 Zero Waste in Production

In the fashion industry's production processes, starting from designing to packaging, the industry is associated with significant resource consumption and material wastage. From excessive fabric cut-offs/ utilization to unsold inventories and harmful finishing techniques, waste is accumulated in each stage of the production cycle. "Zero waste fashion focuses on suppressing the remainder of textile waste or left-over fabric produced in the garment production, which is also known as pre-consumption waste" [14]. It is not about managing waste at the end, but about designing more efficiently and mainly following circular economy principles. This involves a shift from a linear to a streamlined system that considers and values every resource used in the processes. It includes optimizing fabric usage, reducing excess inventory, managing overproduction, minimizing hazardous chemicals, and integrating innovative technologies.

4.1 Fabric Utilization Optimization

In Fashion production, fabric is not only the primary raw material but also the most significant cost and waste contributor overall. "In apparel supply chains, fabric is the single largest material in the cost of a garment; approximately 50–60% of the manufacturing cost can be attributed to fabric alone" [15]. In general, fabric utilization optimization refers to the strategic and technical processes involved in maximizing the use of available fabric during garment production to minimize leftover scraps and reduce pre-consumption textile waste. This concept plays a crucial role in achieving zero-waste goals within manufacturing, especially as sustainability becomes a central concern for the fashion industry.

Optimizing fabric usage not only focuses on simply adjusting cutting layouts, but it also encompasses thoughtful design planning, efficient marker making, innovative software solutions, and collaborative production strategies. To note that, in an environment where even a small percentage of unused fabric translates into tons of waste, fabric utilization optimization is no longer just an efficiency metric but also an ethical environmental imperative.

4.1.1 COP—Cut Order Planning

Cut order planning is defined as deciding how to divide and lay out fabric for cutting garments most economically and efficiently. It is based on various factors such as:

- Order quantity per size and colour
- Fabric width and roll length
- Marker efficiency
- Fabric type
- Machine and labour capacity

The primary goal is to minimize fabric wastage and production costs while meeting the exact quantity and size requirements of orders. The first key component of COP (cut order planning) is order analysis, which helps to assess the total quantity of orders, sizes, and colours required for a particular garment order. COP groups similar requirements into cut blocks (the basic unit for cutting), which in turn saves fabric, time, cost, and labour. The second key component is the marker planning. In general, marker planning is done by placing patterns to use as much fabric area as possible, minimizing the gaps and irregular scraps. But in COP, different cut blocks use different markers to directly improve the marker efficiency of total fabric usage. The third key component is the lay planning, lay refers to the number of fabric layers (piles) stacked to cut multiple garment components in one go. COP decides how many piles of fabric should be used per cut block, depending upon the fabric roll availability, cutting machine height capacity, and order quantity. Balancing the lay height improves time and energy efficiency. The fourth key component is fabric roll allocation, as rolls come in different lengths and widths, they should be assigned wisely. For instance, if the fabric roll is directional or has prints/stripes, the planning should be more complex by matching the rolls to the right lay plan to avoid mismatching of components' prints and roll leftovers.

The apparel industry demands to actively work upon clients' demand, but at minimal cost and wastage. To ensure efficiency and achieve desired outcomes, apparel manufacturers need to streamline key stages of the production process, which include cut order planning, fabric layout optimisation, cutting operations, and balancing production lines. Firstly, cut order planning is all about figuring out the most cost-effective way to arrange pattern sizes on a piece of fabric that has a fixed width. The main goal is to meet the customer's order requirement while keeping fabric waste and costs as low as possible. Then comes the second stage, that is, template design layout, also known as TDL. Here, it involves choosing the most effective way to layout the pattern pieces planned during the COP stage. Once the layout is ready, the third stage involves cutting the fabric using laser cutters. The cut pieces are then grouped into lots and distributed across the production lines for a go-through sewing, assembly, packaging, and then shipped out. In the first stage, cut order planning (COP) is handled either by experienced professionals or specialised software, using an estimated layout length based on various size combinations. The estimation helps to calculate the utilisation of materials. Once COP is completed, the plan moves to the two-dimensional layout (TDL) phase, where patterns are arranged on fabric with fixed width and variable length, aiming to minimise overall production cost and material waste [16].

4.1.2 Hybrid Heuristics

Hybrid Heuristic refers to a problem-solving approach that uses practical methods or rules of thumb (it is a general guideline or practical principle based on experience rather than a scientific formula) to find a good-enough solution quickly. In the context of COP, it helps to make efficient cutting and planning decisions without even trying a single possible outcome (which is commonly done and would consume more time). Heuristic methods estimate and approximate the best layout or grouping by following logical shortcuts.

Different Techniques:

- Greedy Heuristic: prefers the best option at each step (e.g., fill marker space with larger pattern first).

- Constructive Heuristic: one block at a time, building a complete solution from scratch.
- Local Heuristic: starts with one solution and keeps tweaking it to improve marker efficiency.
- Hybrid Heuristics: combine multiple strategies.

Optimising fabric usage isn't about cutting costs, it's about creating a more sustainable fashion industry. Since fabric accounts for a major share of material expenses and pre-consumer waste, managing it wisely is essential. With thoughtful planning, smart technology and precision at every step from cut order planning to marker efficiency and hybrid heuristics, brands can significantly reduce more amount of fabric waste.

Digital tools and intelligent planning system make it easier for manufacturers to meet production targets while using less material, supporting both efficiency and industry's growing commitment to environmental responsibility. When fabric is treated as a valuable resource to be optimised rather than a simple use, fashion companies move close to achieving zero-waste goals without sacrificing profit in today's fast-paced market.

4.2 Modular and Multi-functional Garment

In the journey towards zero-waste fashion, modular and multifunctional garments offer a smart and forward-thinking solution. These designs either serve multiple purposes within a single piece (which is multifunctional) or can be reconfigured using detachable or interchangeable parts (known as modular), giving wearers several ways to style or use them. This approach not only gets more value out of each garment but also encourages mindful buying and offering variety without the need for mass production.

From a manufacturing perspective, using standardised modular components can improve production, boost fabric efficiency and reduce the number of unique SKUs (stock keeping unit). Altogether, modular multifunction garment design supports both sustainable production and consumption by extending the life of clothing, improving versatility and cutting down textile waste starting right from the design table.

4.2.1 Modular Garments

Modular and multifunctional clothing both aim to enhance sustainability and usability, but they differ in approach and design intention. Modular clothing is built from detachable components such as sleeves, collars, and panels that can be interchanged or reassembled, allowing the wearer to customise, repair, and update the garment as needed. This concept is closely aligned with modular manufacturing, where parts are produced separately and assembled later. In contrast, multifunctional clothing is designed to serve multiple purposes or transform within a single piece without needing to detach any components, focusing more on adaptability and convenience. For example, a jacket that turns into a backpack, both of them aim to reduce consumption and extend garment life. Modular clothing emphasises customisation and reconfiguration, whereas multifunctional clothing focuses on utility and performance within a unified form.

Specifically, modular manufacturing is a production approach where a product is divided into separate parts, with each part manufactured independently and later assembled to form the final product. These parts can be made within the same factory or across different locations, even globally. This method has shown proven benefits in the industry, such as a 10% increase in productivity, a 15% boost in labour efficiency, and improved employee morale due to teen waste support, communication, and dignity in the workplace [17]. Successful implementation heavily depends upon management commitment to change, with education and training playing a crucial role in helping employees adapt rather than inspecting quality after production. Modern systems embed quality during design by focusing on customer needs, often using tools like quality function deployment (QFD). However, adopting modularity in any industry requires careful feasibility study and internal alignment to ensure long-term success.

4.2.2 Multifunctional Garment

Multifunctional design in clothing enhances user engagement by allowing garments to adapt to various situations and needs. These designs often include features for assembling or disassembling, offering versatility and comfort. They help extend a garment's life cycle and support eco-friendly fashion practices. This approach balances functionality, style, and sustainability. The need for multifunctional convertible garments arises from the growing demand among females for unique, stylish, and purpose-appropriate clothing. Women are highly experimental with fashion and often avoid repeating outfits for different occasions, leading to unnecessary spending and fabric wastage. Multifunctional garments offer the ability to redesign and restyle the same outfit in various ways, reducing both cost and waste.

Additionally, to move towards a zero-waste approach in fashion, multifunctional garments play a crucial role. Traditional fashion habits, like buying new outfits for every occasion and avoiding repeats, lead to excessive fabric wastage and overconsumption. By allowing a single multiple style and setting, multifunctional convertible clothing reduces the need for frequent purchases. This adaptability minimises fabric usage, extends the garment, life cycle, and supports more responsible consumption. In this way, a multifunctional design alliance with sustainable practices promotes resource efficiency and reduces overall textile waste. If an article of clothing can change colours or be modified by the user in pattern and style, it has a better chance of being worn more often. As a result, a consumer's need for buying and consuming new, different garments is more likely to be reduced [18].

Classification of Multifunctional Clothing

a. Convertible clothing

Garments that physically transform into another type of cloth, with a different structure and look.

E.g., zip-off pants that can be turned into shorts, jackets that can be turned into bags.

5. Reversible clothing:

Designed to be worn inside out or front to back, it offers multiple appearances, gives a new look each time.

E.g., a jacket with two distinct sides, a dress with two wearable orientations.

6. Adjustable clothing:

Garments that can be resized or reshaped to suit different body types, even if there is a change of shape or size, still can be used by adjusting.

E.g., drawstring at waist, adjustable strap, expandable panels.

Advantages of Multi-functional Garments

Multifunctional garments offer a range of advantages, like practical, economical, and environmental benefits, making them an innovative solution in modern fashion. One of the most notable advantages, according to the fashion industry, is reducing over consumption and from the aspect of customer end is space saving/efficiency. A single garment that serves multiple styling purposes significantly reduces storage needs, making it ideal for compact wardrobes. Similarly, they support lightweight, travel, alloy, and individuals to pack fewer items without compromising on variety, which is especially useful for various vacations or business trips, where access to baggage is a concern. This is why multifunctional garments can play an important role in zero waste, as customers would prefer this option more because it is so beneficial in many aspects.

These garments also offer versatility in styling, enabling us to quickly transition from formal office where to party attire or from traditional to contemporary looks just with few adjustments. This adaptability helps avoid the awkwardness of outfit reputation at social events by offering fresh styles from a single base garment.

Constructed with high-quality materials and craftsmanship, multifunctional garments are designed to maintain shape and durability, even when worn inside out or reconfigured, ensuring both functions and aesthetics. They also align with eco-conscious fashion, as the reduced fabric, waste, and saved resources minimise the environmental impact of clothing production. Economically, they support sustainable consumption following the principle of reuse, recycle, and recovery, making them a smart investment for the modern consumer.

4.3 Dead Stock and Overproduction Management

One of the main root causes of environmental waste during the production process in the apparel industry is dead stock. Dead stock in the apparel industry refers to unsold or unused goods that remain after the production or sales cycle. This includes both finished and leftover materials such as fabric remnants, trims, and accessories that were never utilised by manufacturers or sold by retailers. Dead stock fabric, in particular, often arises from cancelled orders, overproduction, wrong-cut textiles, minor defects, or excess cutting waste during garment manufacturing. As these materials accumulate, they contribute to both economic loss and environmental waste. They highlight the importance of finding sustainable solutions to repurpose or reduce deadstock in fashion supply chains.

4.3.1 Causes

The deadstock in the fashion and apparel sector arises from a combination of production inefficiencies and market miscalculations. Common causes include errors in dyeing or printing, excessive ordering by brands, overproduction by textile mills, and material rejections due to quality issues. In retail, it builds up through customer returns, outdated or damaged goods, sizing errors, failed designs, and overstocks from time-sensitive campaigns. Fast fashion intensifies the issue by encouraging mass production and rapid turnover, often leaving a large amount of fabric unused. Alarmingly, around 15% of textiles produced go to waste, translating to an estimated USD 120 billion worth of materials, discarded or stored unused annually [19]. Deadstock storage further leads to quality degradation, especially for natural fibres like cotton and delicate materials, such as lace and elastic. From a business perspective, dead stock ties up capital, increases storage costs, reduces cash flow, and creates challenges in inventory analysis. Therefore, managing overproduction and finding sustainable uses for that stock is essential for both environmental and economic resilience in the fashion industry.

4.3.2 Overproduction Leads to Deadstock

Overproduction is a major contributor to the growing problem of deadstock in the fashion and apparel industry. The fast fashion model, which pushes frequent product releases and short trend cycles, encourages manufacturers to produce large volumes of garments in anticipation of consumer demand. However, when this demand is overestimated or shifts rapidly, it results in excessive unsold inventory. These unsold garments, often stored in the warehouse or retail shelves, become deadstock. It holds no immediate commercial value over time. This debt stock either deteriorates due to poor storage conditions or becomes obsolete due to changing fashion trends. In many cases, these items are incinerated, landfilled, or heavily discounted, contributing to both economic losses and environmental damage that production not only wastes resources like fabric, water, and energy, but also directly fuels the accumulation of dead stock in the supply chain.

4.3.3 Solution Approaches for Deadstock Management

To tackle the growing issue of dead stock in the fashion industry, several innovative strategies and technologies have been adopted. Platforms like Queen of Raw help prevent overproduction by enabling brands to upload and track their surplus fabrics, offering a digital marketplace that promotes circularity. Designers are increasingly using deadstock materials to create new collections, extending the life cycle of unused textiles. At the same time, D2C fashion brands are employing proactive methods such as data-driven forecasting, Limited-edition runs, pre-orders, and agile inventory systems to reduce excess stock. On a technological front, a data analytics model has been proposed that uses decision trees, classifiers, and customer purchase data to predict which items may become deadstock. This system then recommends bundling strategies to pair slower-moving items with best-sellers, thereby optimising sales and reducing waste. These interventions demonstrate that a combination of digital tools, predictive analytics, and sustainable design choices can significantly mitigate deadstock challenges in the apparel industry.

One effective solution to manage deadstock in the apparel industry is the application of data analytics. A research model proposes a smart system that helps classify and clear the stock efficiently. The system works in two main steps: first, it analyses inventory data using machine learning algorithms, like ID3 decision trees and Ada boost to classify products as either "fast-moving" or "deadstock" based on features, such as product style, customer rating, price, and season. This classification helps businesses identify which products are underperforming.

In the second step, the system uses optimisation techniques to create bundles by pairing deadstock items with popular products. By calculating the best combination using linear programming and probability-based algorithms, it recommends bundle discounts that attract customers and help slow-selling stock. This approach not only clears debt stock but also maximizes profit and reduces unnecessary waste, offering an innovative and sustainable solution for inventory management in the fashion industry using data analytics.

4.4 Upcycling and Repurposing Within the Post-production Waste

The textile industry is one of the most resource-intensive and polluting sectors globally, generating significant waste throughout its supply chain. The rise of fast fashion and the shrinking life span of garments have led to a significant search for post-consumer textile waste. While this waste was once seen as a problem to be dealt with, it's now being reimagined as a valuable resource, especially within the framework of a circular economy. Practices like upcycling and repairing machines up becoming key strategies to recover materials and energy from discarded clothing. By giving old textiles new life, the supporters reduce the need for new resources and help lower the environmental impact of fashion, including its carbon footprint [20].

Within this framework, innovative waste management strategies have emerged to align with CE principles. Upcycling means transforming waste materials into products of higher value, while reusing extends the lifecycle of items by giving them a new function. These approaches help reduce landfill use, lower greenhouse gas emissions, and improve economic opportunities through sustainable product development. This section explores two major avenues of such innovation: the upcycling of textile waste into non-woven and composite materials, and the fibre-level recovery and the repurposing of blended fabrics for secondary use.

4.4.1 Upcycling into Nonwoven and Composite Materials

A prominent method for cycling textile waste involves converting it into nonwoven fabrics or fibre-reinforced composites. The process begins by mechanically shredding waste textiles into fibrous form, which is then processed using the needle punching method to create durable, binder-free, non-woven materials. These are utilised in various applications such as wipes, napkins, insulation, and filtration systems [20].

Additionally, textile fibres can be integrated with synthetic or natural polymers to produce composite materials. These materials are not only environmentally friendly but also cost-effective, serving as alternatives to wood-based panels and fibreglass. For instance, recycled denim is used. The fibres have been successfully used in the manufacturing of geotextiles for road reinforcement and building insulation, replacing more resource-intensive construction materials [20]. This method aligns with circular economy goals by reducing reliance on virgin materials, preserving forests, and decreasing industrial emissions.

4.4.2 Recovery and Repurposing of Fibers and Blended Fabrics

Upcycling also involves the mechanical or chemical recovery of fibres from the textile waste, especially valuable in handling blended fabrics. Cotton and synthetic blends, common in post-consumer waste, can be separated through enzymatic hydrolysis or solvent-based methods. For example, cotton-polyester blends can be treated to isolate polyester while converting cotton into glucose, which can then be used to produce bio-products like ethanol or succinic acid [20].

Recovered fibres are often spun into yarns, sometimes mixed with virgin fibres for it to be used in new textile productions. Studies have shown that in cooperating, even 50% recovered cotton into yarn, reduces manufacturing cost significantly while maintaining functional properties, suitable for denim and other fabrics [20]. This approach not only cuts down waste but also alleviates the environmental strain of cotton farming and petroleum-based fibre production. The efficient repurposing of such waste supports both sustainability and economic feasibility in textile manufacturing.

5 Zero Waste in Disposal

The zero waste concept in terms of disposal is to eliminate all types of textile waste that have been generated in the production and consumption lifecycle. Worldwide, 75% of textile waste is landfilled while only 25% is recycled or reused [21]. The final proposal for disposal is to compost all types of textile waste rather than sending them into landfills or incinerators. This concept also shares the ideas of circularity.

5.1 Composting Textiles: Biodegradable Innovations

The two major environmental problems are chemical fertilisers and the production of textile waste [22]. When the disposal of textiles becomes zero, the environmental impacts get reduced, which makes a better natural cycle. One of the biodegradable innovations is composting textiles. Composting textiles is one of the sustainable disposal methods that helps in the breakdown of natural fibers into nutrient-rich soil in the way of biological decomposition. The composting process undergoes sorting, shredding, mixing in the ratio of carbon–nitrogen ratio, and maintaining the conditions of oxygen, moisture of 40–60%, and temperature of 55–65 degree Celsius. The benefits of this process are to divert waste from landfills and encourage sustainable fashion.

5.2 Disassembly Design for Easy Recycling

The global demand has been focused on reducing the expenses of energy and raw materials in the manufacturing process, now become an emerging issue. The result, the market demands for economic recycling and reducing the product and environmental impact at the End of Life (EOL) has increased currently [23]. So, disassembly becomes a very important option for EOL products. Disassembly can improve the life cycle behaviour of products, maintenance, and repair during service.

5.2.1 Smart Textiles Design for Disassembly

Smart textiles combine computerized and textile technologies to prepare smart garments like soft medical devices, space suits, etc., but this smart textile also produces a lot of waste in digital electronics and textiles. The design for disassembly helps to recycle and reduce the impact on the product and the environment. The textile production is one of the most emerging waste and electronics industries globally, producing nearly 50 million metric tons of waste [24]. In the way of considering that smart textiles also support disassembly and recycling, the idea is different from traditional electronic manufacturing.

5.3 Material Recovery Networks

The material recovery networks (MRNs) mainly focus on post-consumer textile disposal, which happens after the product reaches the end of its life from the consumer's perspective. The consumer textile waste is managed to avoid landfills and incineration. Incineration of disposal into MRNs is to take a step in household waste diversion, collecting bins in the neighbourhoods, and taking back programs by brands like H&M and Levi's. One of the best options is reusable and recyclable items. Developing technologies that help in tracing materials post-disposal to check product is routed correctly. By following the circular disposal strategies, which encourage MRNs in consumer education and produce deposit return schemes.

6 Zero Waste in Circularity

According to resources, the fashion industry is the third most polluting industry to pollute the world. Circularity is one of the solutions, and it is the most preferred solution. Nowadays, the circular economy has been welcomed into the fashion industry to provide fast and secure sustainability. The cycle of disposal arises mostly from fast fashion, To avoid this disposal approach, the circular economy strategy has been brought [25].

6.1 Cradle to Cradle Design Philosophy

Cradle to cradle (CTC) in textiles aims to create a circular economy for fabrics. This provides either safer material in return to the environment as a biological nutrient or reusing the new products as a technical nutrient. The goal is to eliminate the waste, and an idea of the circular economy using different methodologies proposed below:

- Cradle to Cradle—The products are continuously reused as raw materials without compromise on quality.
- Cradle to Gate—This speaks about the extraction of resources.
- Gate to Gate—The inputs and outputs of the manufacturing processes are considered.
- Cradle to Grave—This considers the total product life cycle. Includes the extraction of raw materials and the processing of materials for production. The art technology developers are preferred to choose cradle to grave fashion [26].

6.1.1 Biological Cycle and Technical Cycle

In a circular economy, the biological cycle majorly focuses on biodegradable materials, returning them to nature by composting. This cycle allows the land to regenerate nutrients, which produce new biodegradable materials and thus, the cycle closes. The technical cycle mainly focuses on non-generation materials, where the products can become raw materials to produce another product with the same or better features. Non-renewable materials are considered technical nutrients. This cycle gives a safer circulation and continuous usage in the textile industry. The cradle-cradle gives an eco-effective recycling approach to reuse in higher quality [27]. The cradle-cradle concept follows a closed-loop system.

6.2 Closed-Loop Supply Chains in Fashion

The apparel industry consumes a large number of resources, which leads to pollution of the environment. One of the major resources is due to the fast fashion concept. Integrating the fashion features, a system is developed for sustainable fashion products, which is a rent-based closed-loop supply chain [28]. In the rent-based closed-loop supply chain, the products are manufactured according to the required design. The design description should be properly investigated in the markets. The details should be researched according to the target market. The products can be distributed by two methods: traditional retailers and rental services. With the rental services, you can get fashion products online or offline. The disposal can be done by consumers in a distributed manner, or the suppliers can dispose of it in the proper manner [28]. Closed-loop supply chains have been implemented by many apparel brands.

6.2.1 Apparel Brands

H&M

H&M follows the circular economy. According to data, less than 1% of materials are used for recycling in the apparel industry. This makes us understand that thousands of tons of textiles end up in landfills. This made the H&M [29] brand move towards reusing or recycling fashion. The brand brought a collecting program in 2013, which was the world's biggest campaign. They got unwanted clothes and also provided thank-you vouchers to welcome the customers. In 2019, the brand launched a rental service in their store in Stockholm, where garments can be rented from their conscious exclusive collections.

Levi's

In 2020, they launched a program that supports closed-loop supply chain systems. The main aim is to recycle the old Levi's into new products. By doing this, they were achieving an impact on cotton agriculture, where the denim is brought from the consumer's closets. The brand brought an idea of separating cotton-made denim from polyester blends so that it can be recycled easily. The brand has collaborated with its network of suppliers to achieve sustainable apparel conditions.

6.3 Garment-to-Garment Regeneration Technologies

The garment-to-garment regeneration refers to the recycling systems of textiles that are used for transforming old or unwanted garments into new clothing without returning them as raw fiber. These technologies mainly focus on creating a closed-loop system in the apparel industry. The technology is divided into various types:

- Mechanical fiber regeneration—In this process, the old garments are shredded, cleaned, and a mechanical process is done to respun the yarns.
- Chemical fiber regeneration—This is to use solvents to break down the garments into base polymers or cellulose, which helps in the regeneration of new fibers.
- On-demand micro factories—Small-scale production systems, which provide immediate regeneration.
- Thermal regeneration—The synthetic fibers can be re-extruded like polyester.

6.4 Community Centre in a Circular Fashion

Community centre in circular fashion is based on the community, where it is treated equally in setting a process or a goal [30]. In the community center, as a hub to engage the consumers in a sustainable fashion practice, garment regeneration helps grow circular fashion. The major community hubs are the upcycling studio, the fashion library where the garments can be borrowed, the microfactory, and circular startups. Education and awareness programs can also be provided in film screenings, and some useful workshops on textile waste can be held at a community center. Fabscrap is a textile reuse or recycling center where there are volunteers to work on textile waste in landfills. Nine years in, two million pounds of textiles have been diverted from landfills [31]. These types of community centers help in making sustainable fashion in real life.

7 Future Scope of Zero Waste Approach

Zero waste design offers a great level of future scope:

- Brands' adoption of zero waste design brings a huge difference in textile waste and protects the environment. To support this statement, a brand named 'I was a sari' gave a thought of upcycling saris in Mumbai, India. By adopting this type of approach, they collected about 15–20 tonnes per day.
- Making technology advancements in pattern design by introducing 3D modeling &AI tools and customizable modular patterns for waste-free alterations.
- Following the circular economy in the form of closed-loop systems gives a designed product for reuse, repair, and the process of upcycling. A community in Guwahati brought the idea of upcycling the discarded clothes into handbags, cushion covers, and promoting sustainable craftsmanship.
- Moving towards biodegradable or recyclable materials helps in reducing textile waste.
- Creating patterns for zero waste garments helps consumers to build concern for their clothing into sustainability.
- Providing transparency in labelling gives an organic trade label, which becomes a fair selling point.
- Government policies towards the reduction of textile waste support the zero waste design concept. Bharat Tex events have helped to promote textile upcycling across India.
- In the educational curriculum of designing schools, they should have zero waste design concept as their mandatory subject to improve the thought of new designers towards sustainability.
- A study indicated the concerns on utilisation of pre-consumer textile waste and how these wastes can be converted to designing children's clothing wearables, thereby minimising waste [32].

8 Conclusion

Zero waste design is a concept of developing sustainable fashion. The key aim is to eliminate textile waste at the start of the design stage to avoid managing the process after production. Taking into account the thoughts of traditional pattern making, also including technologies like laser cutting and digital modeling, these concepts help us in minimizing waste and provide creativity to the young designers. The fashion has moved towards circularity, and zero waste design plays an important role in reuse and recycling systems. Introducing technologies like 3D visualization and AI-assisted pattern generation gives a dynamic shape to the apparel industry. Apart from environmental impacts, zero waste design provides social and economic implications. The zero waste design is not only termed as a concept, but it also provides respect towards the materials, laborers, and the environment. In India, apparel industries reflect the blend of traditional craft and modern sustainability practices and have looked at all avenues to minimise the wastage. Many brands have proved that fashion can be both eco-friendly and commercially viable. The apparel industries have incorporated zero waste design measures leading to more environmental conservation. The conservation includes managing deadstock in the apparel industry, utilising pre-consumer textile waste to create children's wear, ensuring zero waste in pattern making, making design variations like dart manipulation, modular designing, and modular manufacturing. On the sides of consumers or creators, the zero waste design should be increased in the value of sustainability. The concept creates a transformative pathway towards a responsible and regenerative apparel industry.

References

1. Grewal D, Guru R, Kumar A, Arora N (2023) To study on zero wastage in pattern making process. In: International Conference on Sustainable Design Practices, NIFT (Kangra), India
2. Liu M (2015) Fashioning geometric patterns: investigating the underlying geometry of fashion patternmaking (Doctoral dissertation, University of Technology Sydney (Australia))
3. Rissanen T, McQuillan H (2023) Zero waste fashion design. Bloomsbury Publishing
4. Rachael D, Jackie G (2022) The global glut of clothing is an environmental crisis. [online] https://www.bloomberg.com/graphics/2022-fashion-industry-environmental-impact/. Accessed on 10th May 2025
5. Papahristou E, Bilalis N (2016) A new sustainable product development model in apparel based on 3D technologies for virtual proper fit. In: Sustainable design and manufacturing 2016. Springer International Publishing, pp 85–95
6. Fletcher K (2013) Sustainable fashion and textiles: design journeys. Routledge
7. McKinney EC, Bye E, LaBat K (2012) Building patternmaking theory: a case study of published patternmaking practices for pants. Int J Fash Desig Technol Educ 5(3):153–167
8. Xu J, Zhang X, Wang S, Wu J (2013) Tool path generation for pattern sculpting on free-form surfaces. Int J Adv Manuf Technol 67:2469–2476
9. Yuan W, Liu Y, Zhao J, Wang H (2016) Pattern-based integration of system optimization in mechatronic system design. Adv Eng Softw 98:23–37
10. Roberts J (2021) Subtraction Cutting 2014–2020–131 Scholarly Encounters and Artefacts [REF 2021 Portfolio]
11. McQuillan H (2020) Digital 3D design as a tool for augmenting zero-waste fashion design practice. Int J Fashion Design Technol Educ 13(1):89–100
12. Mansour OSM (2017) Integrating the concept of modular design and dart manipulation technique for the innovation of fashion design for women. Int Design J 7(4):87–97
13. Huck J, Maganga O, Kim Y (1997) Protective overalls: evaluation of garment design and fit. Int J Cloth Sci Technol 9(1):45–61
14. Nursari F, Djamal FH (2019) Implementing zero waste fashion in apparel design. In: 6th Bandung creative movement 2019. Telkom University, pp 98–104

15. Wong WK, Leung SYS (2008) Genetic optimization of fabric utilization in apparel manufacturing. Int J Prod Econ 114(1):376–387
16. M'Hallah R, Bouziri A (2016) Heuristics for the combined cut order planning two-dimensional layout problem in the apparel industry. Int Trans Oper Res 23(1–2):321–353
17. Sudarshan B, Rao DN (2013) Modular manufacturing in the garment industry. Int J Eng Sci Innov Technol 2(5):219–230
18. Black S (2011) Sustainable design strategies: eco chic, the fashion paradox. Text: J Textile Soc 38:24–30
19. Banikya AK, Ranade P, Verma IK (2024) Managing deadstock in the fashion and apparel industry: an exploration of causes, solutions, and technological interventions. Eng Proceed 66(1):9
20. Stanescu MD (2021) State of the art of post-consumer textile waste upcycling to reach the zero waste milestone. Environ Sci Pollut Res 28(12):14253–14270
21. Juanga-Labayen JP, Labayen IV, Yuan Q (2022) A review on textile recycling practices and challenges. Textiles 2(1):174–188
22. Biyada S, Merzouki M, Imtara H, Benlemlih M (2019) Assessment of the maturity of textile waste compost and their capacity of fertilization. Eur J Sci Res 154:399–412
23. Abuzied H, Senbel H, Awad M, Abbas A (2020) A review of advances in design for disassembly with active disassembly applications. Eng Sci Technol Int J 23(3):618–624
24. Wu S, Devendorf L (2020) Unfabricate: designing smart textiles for disassembly. In: Proceedings of the 2020 CHI conference on human factors in computing systems, pp 1–14
25. Papamichael I, Voukkali I, Loizia P, Rodrıguez-Espinosa T, Pedreño JN, Zorpas AA (2023) Textile waste in the concept of circularity. Sustain Chem Pharm 32:100993
26. Zhang L, Leung MY, Boriskina S, Tao X (2023) Advancing life cycle sustainability of textiles through technological innovations. Nat Sustain 6(3):243–253
27. El Haggar S (2010) Sustainable industrial design and waste management: cradle-to-cradle for sustainable development. Academic Press
28. Hu ZH, Li Q, Chen XJ, Wang YF (2014) Sustainable rent-based closed-loop supply chain for fashion products. Sustainability 6(10):7063–7088
29. H&M (2025) 'Closed loop campaign',[Online], https://www2.hm.com/en_in/sustainability-at-hm/our-work/close-the-loop.html?srsltid=AfmBOorC4fRN8t2xK3VJD46EbWg-uhp0Nu731Z7YF8oMMIuQUR5kw9uQ. Accessed on 12 May 2025
30. Handayani RB (2023) November. fashioning a greener future: teaching circular fashion and engaging Surabaya's zero waste community. In: International conference on sustainability in creative industries. Cham, Springer Nature Switzerland, pp 67–76
31. Nicolaus T (2025) Textile waste a growing issue as Americans consume more [Online], https://www.thenationshealth.org/content/55/3/1.1?utm. Accessed on 23rd August 2025.
32. Bhown H (2024) Design and development of children's wear using pre-consumer textile waste: a sustainable approach.[Online]. https://textilevaluechain.in/in-depth-analysis/articles/research-paper/design-and-development-of-childrens-wear-using-pre-consumer-textile-waste. Accessed on 23rd August 2025

Sustainable Solutions in Textile and Fashion Industry

Gizem Karakan Günaydın and Ozan Avinc

Abstract The fashion sector is one of the largest providers to environmental pollution and social inequality, making sustainability a crucial issue. The rise of fast fashion has resulted in excessive resource consumption, increasing carbon emissions, and growing textile waste. High water usage, chemical pollution, and energy-intensive production processes significantly impact the environment, while unethical labour conditions raise serious concerns about social sustainability. With the growing awareness of sustainability in fashion, the adoption of eco-friendly materials and innovative production techniques has gained momentum. Various sustainable solutions have emerged, including eco-friendly materials, innovative production techniques, and circular business models. Nowadays, fashion brands are increasingly adopting sustainable practices to minimize their environmental impact and promote ethical production. Many global brands have committed to reducing their carbon footprint by using eco-friendly materials such as organic cotton, recycled polyester, and bio-based fabrics. Additionally, innovative production techniques like waterless dyeing, 3D knitting, and digital fashion are being integrated to reduce waste and resource consumption. Another major shift is the adoption of circular economy models, where brands promote clothing rental, resale, and take-back programs to extend the lifecycle of garments. Companies such as Patagonia, Levi's, and H&M have introduced initiatives that encourage customers to return old clothes for recycling or resale. Some brands are also experimenting with biodegradable textiles and regenerative farming practices to create a more sustainable supply chain. This chapter will cover key strategies such as the use of organic and recycled fibres, low-impact dyeing processes, 3D knitting, and digital fashion. Furthermore, concepts like slow fashion, zero-waste design, and upcycling are discussed, illustrating how brands and consumers can contribute to sustainability. Moreover, the role of emerging technologies such as

G. K. Günaydın
Faculty of Architecture and Design, Textile and Fashion Design Department, Pamukkale University, Denizli, Türkiye

O. Avinc (✉)
Faculty of Engineering, Textile Engineering Department, Pamukkale University, Denizli, Türkiye
e-mail: oavinc@pau.edu.tr

S. S. Muthu (ed.), *Sustainable Solutions in Textiles and Fashion*, Sustainable Textiles: Production, Processing, Manufacturing & Chemistry,
https://doi.org/10.1007/978-3-032-10925-5_6

artificial intelligence, blockchain, and digitalization in augmenting transparency and sustainability are also be discovered.

Keywords Sustainable fashion · Ecofriendly materials · Carbon footprint reduction · Digitalization

1 Introduction

Sustainable fashion is a comprehensive approach focused at minimizing the social and environmental impacts of the textile and fashion industries. It involves the use of eco-friendly materials and production techniques, the promotion of responsible consumer behavior, and the advocacy for fair labour practices. In contrast to conventional fashion which often prioritizes speed, low cost, and high-volume sustainable fashion seeks to reduce the ecological footprint throughout the complete life cycle of a garment. This includes the thoughtful selection of organic and recycled materials, the reduction of CO_2 emissions during production, and a commitment to transparency across the supply chain.

Beyond material choices, sustainable fashion embodies a transformative lifestyle philosophy, embracing the principle that "less is more." It emphasizes durability, craftsmanship, and mindful consumption over fleeting trends and disposable clothing. By opting for sustainably produced garments, consumers can help preserve natural resources, reduce environmental pollution, and support better working conditions and livelihoods for those employed in the textile sector.

At its core, sustainable fashion is rooted in the responsible stewardship of natural resources. It favors materials like organic cotton, which have a considerably lower environmental impact than synthetic alternatives. Moreover, it encourages the use of low-impact production techniques aimed at reducing the industry's greenhouse gas emissions and other environmental harms. Understanding sustainable fashion also requires consumer awareness of brand transparency in sourcing and production practices. Truly sustainable brands often carry globally recognized certifications, such as the Global Organic Textile Standard (GOTS) and Fair Trade, which validate their commitment to ethical labour standards and environmental responsibility.

Ultimately, sustainable fashion represents a lifestyle that respects planetary boundaries and prioritizes long-term ecological health. Supporting this movement empowers individuals to contribute to the creation of a more ethical, socially just, and environmentally sustainable fashion industry for both present and future generations [1].

There are some examples of clothing trades focused on sustainability; Jeanologia is recognized as a pioneer in sustainable innovation, revolutionizing traditional, water-intensive denim finishing methods by introducing advanced laser and ozone gas technologies. These techniques significantly reduce water usage and environmental impact, while also eliminating the need for harmful chemicals commonly used in conventional processes [2]. Palomo Spain exemplifies ethical and sustainable fashion practices by prioritizing small-scale, local production. The brand's commitment to sourcing sustainable materials and embracing artisanal craftsmanship promotes a more thoughtful and environmentally responsible model of fashion consumption [3].

In a similar vein, artisans from Kyrgyzstan highlight the value of sustainable tailoring by utilizing traditional felting techniques. Their small-scale production methods rely exclusively on natural materials, helping to preserve cultural heritage while actively contributing to the conservation of natural resources. Together, these examples demonstrate the feasibility and growing adoption of sustainable tailoring within the fashion industry. They underscore a collective commitment—among both brands and independent artisans—to environmental stewardship, cultural preservation, and the promotion of fair labor practices.

This chapter includes some environmental impacts of fashion and textile industry considering the whole life cycle of a product from raw material to disposal. Moreover, sustainable materials, eco-innovations, the possible ways to increase the sustainability, sustainable business methods for textile and fashion, sustainable solutions for some apparel brands and challenges to sustainability in textile and fashion were also discussed. Finally, the role of artificial intelligence on textile and fashion was also discussed.

2 Environmental Impacts of the Fashion and Textile Industry

Over the previous three decades, the quantity of clothing bought per person in the EU has increased by almost 40%. The quickening of fashion cycles and falling prices are largely responsible for this spike. An estimated 2% to 10% of the environmental impact associated with EU consumption is attributable to clothing [4, 5]. However, because a large portion of textile production takes place outside of the EU, the environmental effects are frequently seen outside of it. Growing raw materials, turning them into yarn, weaving, and dying textiles all require a significant amount of chemical and water input, including insecticides for cotton and other crops.

Due to the energy, water, and chemicals needed for washing, drying, and ironing, along with the introduction of microplastics into ecosystems, the garment consumption phase also plays a major role in environmental deterioration. Because the technologies required to regenerate fibres are still in the early phases of development, less than 50% of waste clothing is collected for reuse or recycling, and just 1% is regenerated into new apparel [4, 5].

Many ideas have been put up to overcome these issues. These include introducing cutting-edge business models like clothing rental, advocating for design principles that facilitate simpler recycling and reuse (circular fashion), and urging customers to buy fewer, higher-quality, and more long-lasting clothing items (slow fashion).

In general, consumers must change their habits to make more sustainable decisions [6, 7].

Under comments from the European Parliament, the European Union adopted a circular economy package in 2018 that requires all Member States to undertake separate textile collection by 2025. The Parliament has long backed programs that encourage the reuse and recycling of clothing aside from the utilization of sustainable raw materials [8].

When the environmental impact of textile and fashion industries is considered, it is understood that it includes the whole life cycle of a product from the raw material to disposal. In fact, less than 1% of clothing is recycled and the rest is usually abandoned or incinerated which results with poisoned rivers, reduced soil fertility, climate change. Health risks, child labour, destruction of local economies should also be mentioned as additional results. In the 90's, the term of fast fashion started to be mentioned which referred to a business model with large collections (up to 52 in one year). However, the collections included garments of low-level quality with affordable prices. The fast fashion trend encourages the compulsive buying behaviour of products which are thrown away within a few months. Today, brands such as Zara, H&M, and Forever 21 pioneered this model, offering consumers fashionable styles at low prices with high turnover rates. The goal of fast fashion is to maximize consumer purchasing frequency, fueling a culture of overconsumption [4–7].

Environmental impact of fast fashion should also be considered as specified in below [4–7]:

- Resource Depletion: The industry consumes enormous quantities of water, energy, and raw materials. For instance, the manufacture of a single cotton T-shirt requires approximately 2700 L of water.
- Pollution: Textile dyeing is the second-largest polluter of water globally. Additionally, synthetic fibers like polyester shed microplastics, contaminating oceans and aquatic ecosystems.
- Waste Generation: Because of the low cost and perceived disposability of garments, millions of tons of textile waste end up in landfills each year. Many items are discarded after only a few wears.

3 Sustainable Materials and Eco-Innovations

3.1 Selecting the Right Raw Material

A noteworthy portion of the textile and apparel industry's environmental impact is caused by the manufacturing of raw materials, including the cultivation of crops for natural fibers. Cotton is particularly problematic because cultivation requires enormous amounts of land, water, fertilizer, and pesticides. According to a 2015 assessment by the European Clothing Action Plan (ECAP), cotton makes up over 43% of all fibers used for clothing on the EU market. Because bio-cotton consumes less water and produces fewer pollutants than conventional cotton, its environmental effects can be significantly lessened. A Textile Exchange research states that the percentage of sustainable cotton rose from 6% in 2012–2013 to 19% in 2019–2020 [9].

Silk is particularly harmful with regards to global warming and the depletion of natural resources, cotton is a chief contributor to water scarcity, and wool is a foremost contributor to greenhouse gas (GHG) emissions. But the industry is also experimenting with less common natural fibers that use less water, fertilizer, and pesticide, such hemp, flax, linen, and nettle. Polyester, a non-biodegradable fiber derived from fossil fuels, makes up 16% of the fibers used in clothing [7]. Its primary benefits are a lower water footprint than cotton, the need for lower washing temperatures, quick drying and minimal ironing, and the ability to be recycled into fresh (new) fibers. Made mostly from plastic bottles, recycled polyester raised between 2007 and 2017, increasing its market share from 8 to 14% [7]. But according to a number of recent studies, washing a single load of polyester clothing (together with nylon and acrylic) can release 700,000 microplastic fibers into the environment, which can then enter the human food chain and release poisons. An estimated half a million tons of plastic microfibers from clothing washing are thought to enter the ocean annually. Currently, the industry is experimenting with biobased polyester, sometimes referred to as biosynthetic, which are derived at least partially from renewable resources such plant oils, sugar cane, corn, or beets. The difficulty lies in locating feedstocks that don't interfere with food production and don't need a lot of water or pesticides [7–9].

In the EU, about 9% of the fibers used in clothing are created from manmade cellulosic (MMCs), which are cellulose obtained from dissolving tree pulp. Viscose, sometimes referred to as rayon, is the most widely used [10]. Although they are biodegradable and derived from renewable plants, the primary issue is acquiring cellulose sustainably, since the output of MMCs more than doubled globally between 1990 and 2017. Therefore, the industry is experimenting with more sustainable materials like Piñatex (made from pineapple leaves), Bemberg (also known as cupro, made from cotton linter that cannot be used to spin yarn), and Lyocell (also known as Tencel, a brand name for cellulose from eucalyptus that grows quickly and doesn't require irrigation or pesticides) [10]. The industry is also testing less frequently used natural fibres, such as hemp, flax, linen, and nettle, that require less water, fertilisers and pesticides.

The fibers of *Linum usitatissimum*, a member of the linen family and one of the earliest crops to be farmed for use in textiles and food, are commonly referred to as flax. Flax's diverse properties have been prized for a number of reasons throughout history. Because its fibers were superior to other fibers and enjoyed by the affluent, they could be hand-made into sturdy, ornamental, gossamer-like woven linens and lace, giving it a high position in the fabric hierarchy. Undoubtedly, this was an ancient premium fiber [11]. Ancient peoples may have spun wild flax fibers to make linen-like textiles, as evidenced by the finding of dyed flax fibers in a Georgian cave that dates back to 34,000 BCE. Additionally, flax bandages have been found wrapped around mummified bodies in Egyptian tombs and used in ancient Egyptian medicine [11–20] and the creation, use, and cultivation of linen cloth are detailed in tomb images. Because the flax stalk's center contains long, fine fibers that could reach up to one meter in length, such high-quality cloth was possible before the contemporary industrial age.

Simple hand spinning tools like a distaff and spindle can be used to effortlessly spin delicate flax fibers into incredibly fine thread. Before being put together to create garments and furnishings, the thread may be hand-braided and dyed into fishnets and ropes or woven into strips of fabric on crude looms. The plant has also been grown for its myriad byproducts, which include soap, paper, colors, meals, and medications. Linseed oil is an edible oil made from flaxseed, the seeds of the flax plant. The fiber's many applications, such as the production of luxury textiles, gained prominence and popularity during the Industrial Revolution. However, as oil-based fibers like polyester, which had a comparable handle and strength but were crease-resistant and can be mass-produced at a low cost, began to replace the fiber, its market share and popularity started to decline. Unlike other fibers like cotton and wool, which were marketed by Cotton Incorporated PLC and the International Wool Secretariat, respectively, flax did not establish a brand name in the 1960s [21–31].

It is an accomplishment that flax products have been reintroduced in new forms in the bio-composites industry, for use in construction materials, automobiles, airplanes, and furniture, as well as in the medical field. In addition to pointing out the problems and concerns with these non-biodegradable composites, which are made from non-renewable resources and are using up remaining oil resources, Blackburn [26] highlights the significance and impact of the development of synthetic chemical products for composites used in large quantities of products. Bio-composites based on natural fibers like flax, jute, and hemp are becoming more popular since they are robust, lightweight, corrosion-resistant, and possess decent sound and moisture absorption properties. A material must be able to be "broken down into simpler substances (elements and compounds) by a naturally occurring decomposer" in order to be considered "biodegradable." The entire cradle-to-cradle lifetime of the fiber or product must be taken into account for the material to be considered "bio". Only recently have novel uses for flax fiber come to light. For instance, a wide range of novel composite materials and applications for new luxury premium products with a clear supply and disposal chain from start to finish can be created by combining incredibly diverse bio-composites with natural fibers and/or coating them in different bio-resins [26].

The flax story claims that sustainability is the new luxury and that the business can easily and honestly communicate its benefits to the consumer. New agricultural mechanization and technological developments in the production of fibre, yarn, and textiles have improved and refined the materials for a number of high-value industries. The technical properties of flax were studied and developed to produce high-quality, wear-resistant fabrics free of defects. The hollow fibers created fabrics that were breathable and comfortable in cold weather, making them suitable for installation in any season. This was due to the fibres' ability to absorb 20% of their dry weight in moisture without feeling wet. In the summer, this assisted in controlling body temperature. Flax also has antistatic properties, is non-allergic, antimicrobial, and naturally resistant to mildew and moths (it is utilized in some medical sutures). Since the fiber absorbs and fixes color well, the color enhances and often intensifies as the fabric ages and the internal cores of the fibres show. The desire for eco-friendly textiles has prompted a focus on the development of low-impact reactive dyes and fabric finishes, such as enzyme finishes and wrinkle-resistant treatments, and flax has a preference for colours that use few dyestuffs [13–26].

3.2 Processing and Garment Production

Spinning raw materials into yarns, weaving them into fabrics, and adding finishing treatments like dyeing or giving the fabrics strength and shine are energy-intensive procedures that require large amounts of water and chemicals. The EU considers 165 of the more than 1900 chemicals used in clothing production to be harmful to human health or the environment. Up to 150 L of water can be used for dyeing every kilogram of fabric, according to the 2024 Global Fashion Agenda [31]. Since most production takes place in underdeveloped countries, where environmental rules are less stringent than in the EU, wastewater is routinely dumped unfiltered into rivers. When making clothes, a lot of energy is used by the equipment used for sewing, gluing, welding, and seam taping. The cut-offs that remain after the clothes patterns have been cut out also contribute to over 20% of the fabric waste generated by the company.

The JRC (European Commission, Joint Research Centre) 2014 study on the environmental improvement potential of textiles recommends a number of tactics to reduce the environmental effects of the processing and manufacturing phase [27]. These include recycling water, using dye controllers and water-efficient dyeing equipment, and using fewer chemicals and enzymes instead of chemicals. It also recommends integral knitting, which creates a whole garment in one piece, obviating the necessity for cutting and sewing. However, the lifetime gains can be outweighed by the integral knitting machines' high energy usage. Some businesses are experimenting with different cuts, computer-controlled pattern-making technologies that allow them to utilize more fabric with fewer cut-offs, garments with few or no seams, bonding or gluing instead of sewing, and other cutting-edge dyeing methods. For instance, the Dutch business DyeCoo is testing the use of CO_2 as a dyeing media in place of water [27, 31].

3.3 Distribution and Transportation

Long delivery routes result from the importation of the majority of textile raw materials and finished goods into the EU. Nevertheless, the Global Fashion Agenda report states that this stage only 2% of the industry's climate change effects, since the majority of major businesses have optimized the flow of commodities [31]. But this stage is also marked by waste from tags, hangers, bags, and packing, as well as a significant percentage of goods that are never sold since the unsold inventory is thrown away.

3.4 Consumer Use

According to the European Commission, Joint Research Centre (JRC study), this stage of the clothing lifecycle has the biggest environmental impact because of the water, energy, and chemicals (mainly detergents) utilized for washing, tumble drying, and ironing, together with the microplastics that end up in the water [27]. Therefore, the report came to the conclusion that encouraging consumers to adopt small behavioral changes, like lowering the washing temperature, washing at full load, avoiding tumble-drying, buying eco-friendly fibres, and donating used clothing, is one of the most efficient ways to lessen the environmental impact of clothing. By washing their clothes less regularly and eliminating needless ironing, consumers can further lessen the environmental effect of their clothing [27–31].

3.5 Recycled or Reused

It is difficult to find current, trustworthy information about what happens to clothing after its owners decide to part with it. It appears that the majority of clothing in the EU is still thrown away and either burned in incinerators or ends up in landfills where methane is released. According to estimates from the Textile Recycling Association, which the European Commission, Joint Research Centre (JRC) cited, only 15% to 20% of textiles that were disposed of in 2005 were gathered for recycling or reuse [27]. There are significant disparities among EU nations, according to research conducted by the European Clothing Action Plan (ECAP) that concentrated on six EU Member States [27, 28].

It is uncertain how much of the clothing gathered is non-reusable and what percentage is reusable. After being gathered, clothing can either be recycled or used again as second-hand clothing. Both routes have problems at the moment. In the EU, the supply of used clothing exceeds the demand, and a significant portion is exported, partially to East Asian or African nations. This has led to claims that waste is transported to nations that cannot handle it and that low-cost used clothing contributes to the demise of regional textile businesses. Although discarded textiles might pose a challenge to domestic textile production, research done for the Nordic Council of Ministers found that more affordable new clothes imports from Asia are likely to fill the void. It also came to the conclusion that garbage from European clothing is not actually shipped to developing nations [27–31].

Fewer than 1% of all materials used in clothes are recycled back into apparel worldwide because of a number of problems with recycling [31]. The lack of technologies for sorting the gathered clothing, separating blended fibers (though some promising new technologies, like Worn Again and Blend Re:wind, can separate the most common blend of cotton and polyester), separating fibers from chemicals, including color during recycling, and determining which chemicals were used in the manufacture in the first place (which is one reason it is easier to recycle factory waste like cut-outs) are all reflected in this.

Furthermore, there are currently insufficient technologies that would allow clothing to be recycled into virgin fibers. For this reason, most clothing is recycled mechanically by being cut up and shredded, which results in shorter, lower-quality fibers that lose 75% of their value. As a result, they are typically recycled into mattress filling, wiping cloths, and insulation rather than being utilized to make new clothing. Mechanically recycled cotton can now be combined with polyester or synthetic cellulosic thanks to new technology. For polyester and nylon, there are chemical recycling technologies that yield high-quality virgin fibers; for cotton and mixes, these technologies are gradually emerging but are not yet completely commercially feasible [27–31].

Significant new initiatives are being made today to reduce the waste and environmental risks associated with textile production or to create green and sustainable materials. One of these promising approaches is to promote the usage of SaXcell fibre in textile products as virgin (100%) or with other blends. SaXcell, an abbreviation of Saxion cellulose, is a regenerated virgin textile fibre made from chemically recycled domestic cotton waste. The process to transfer domestic cotton waste into SaXcell fibre is a crucial step in the circular textile chain. SaXcell production starts with sorting domestic cotton textile waste into an as pure as possible, well-defined waste stream. Next, the pure waste stream is grinded and non-textile components like zippers, nails and buttons are removed. The result is a dry mixture of textile fibres with different fibre lengths. All fibre lengths, long and short, are suitable as raw material for SaXcell. The dry mixture exists of different colours and is chemically decoloured and made suitable for the wet spinning process. Wet spinning can be done according Viscose or Lyocell processes. This means that the new fibre can be produced with few adjustments in the settings of machines on existing installations in Europe and Worldwide. The end product of this step is SaXcell, a regenerated virgin cellulose fibre. The fibre can be cut to specified lengths, spun into yarns and woven or knitted into fabrics. Colouration step can be applied to the fibre, yarn, or fabric stages [29].

4 Possible Ways Forward for Textile and Fashion Industry

The industry is pushing hard to make all stages of production more sustainable.

It is known that major fashion brands and sportswear firms are setting the standard for investments in innovative technology and business practices [31].

However, mid-range retailers are also seeing significant advancements, and even quick fashion is becoming more environmentally friendly. There have been warnings that businesses who don't alter their practices risk running out of resources in the future because of the growing cost of commodities. The endeavor is challenging, nevertheless, because, for example, attempting to lessen environmental effects may result in higher pricing for customers, and persuading customers to purchase fewer items of clothing may result in lower earnings for businesses. Finding a more sustainable fabric mix to use less conventional cotton, enhancing sorting and recycling technologies, improving washing and drying efficiency, boosting energy effectiveness and the utilization of renewable energy in technological processes, prolonging the life of clothing, and improving sorting and recycling are some of the recommendations made by several studies [9, 31]. The possible ways to increase the sustainability may be summarized as below.

4.1 Extending Longevity of Clothes

The estimate research reveals that if the number of times of a cloth is worn doubled on the average, the emissions would be 44% lower [9, 31]. Some assumptions were considered for this.

Slow fashion: Slow fashion is an attempt which encourages customers to purchase fewer clothes of better quality and to keep them for longer. The philosophy includes reliance on trusted supply chains, small scale production, traditional crafting techniques, utilizing local materials and trans-seasonal garments. It emphasizes an alteration in the economic model by selling fewer clothes. However, it may threaten the economic survival of garment producers if consumers do not want to pay high prices. Buldan hand-woven clothes are good examples of slow fashion. For ages, cotton has been grown and woven extensively in Turkey's Aegean Region. Denizli city's "Buldan" town is well-known as the weaving hub for the production of the renowned "Buldan cloth.“ Early historical records formally demonstrated that Buldan town supplied a significant portion of the woven fabric required by the Ottoman Royal Palace. Although Buldan fabrics were initially primarily made for home textiles and underwear, their comfortable qualities now allow them to be used for outdoor clothing as well. Because of their highly twisted yarns, those textiles also offer a great degree of flexibility. Buldan residents began making textiles many years ago by weaving various cotton fabrics in their living rooms or basements. Consequently, this production method has made the contemporary Buldan weaving industry not merely a company, but a way of life in Buldan. As a result, the textile industry began in Buldan homes many years ago, making weaving there more than simply a business; it is a means of passing down Buldan culture and customs. Since the nineteenth century, the town's residents have been weaving the renowned traditional fabric known as "Alaca," which is a cotton garment with stripes of various colors. Among the vast array of Buldan textile products, a significant number of handkerchiefs, home textiles including towels and bed linens, vibrant violet silk (pesthemal) to wrap around the body, and vibrant head coverings are also worth mentioning. Buldan continues to create and sell traditional handwoven goods. However, as the town's industrial development extended, semi-automated and sophisticated automatic weaving looms also started to be employed in addition to handwoven textile items. As the town's industrial development has grown, sophisticated automated looms have begun to be used extensively, leading to the export of goods in piece/part and metered weave form to several nations. The production of Buldan handwoven and machine-woven fabrics, as well as the export of textile products made from these fabrics [in the form of finished products like clothing (t-shirt, shirt, short, trousers, dress, scarf, etc.), home textiles (towels, pesthemal, table linen, bed coverings, curtains, etc.), piece goods, and over-the-counter fabrics] to numerous countries, is growing daily due to the environmentally friendly and sustainable nature of handwoven products and the mass production potential of semi-automatic and automatic weaving machines [30]

4.2 *Improved Collection for Re-use Repair and Up-Cycling*

Some brands facilitate this process by selling second-hand clothes in their own stores. Some brands offer long-term guarantees; these include free repairs, providing repair instructions, and services for reshaping or upgrading clothes.

4.3 *Smart and Instant Fashion Approaches*

Global Fashion Agenda report (2024) indicates that smart and instant fashion applications have the potential to significantly reduce waste. Smart fashion predicts that future garments will be equipped with technologies that instantly adapt to consumer preferences (for example, color-changing), thereby eliminating the necessity to create multiple versions of the same clothing [31]. Instant fashion can enable production according to demand at the point of sale; with advanced 3D printing technologies, this production can make local manufacturing within the EU possible. In this way, the consumer can have the product they want produced locally without excessive stock. Some small brands are preventing overproduction by manufacturing according to consumer orders [31].

4.4 *Targeting Consumers*

Efforts to make clothing more sustainable require consumers to make conscious choices and accept these changes. There are numerous approaches that could be applied in this area:

Increasing consumer awareness: By providing information about sustainability in stores or on clothing labels, or through campaigns, consumers can be encouraged to only purchase the products they need and to prefer more sustainable options. Research shows that while consumers generally support environmentally conscious fashion, this attitude does not always reflect in their behavior. For example, the slow fashion movement has not received as much attention as the slow food movement. Among the reasons for avoiding sustainable fashion are higher prices (despite the increasing number of conscious consumers willing to pay more for sustainable options), and the perception that recycled clothing is of low quality or lacking in style.

Increasing transparency and environmental labeling: Some brands provide customers with information about the ecological impacts of their products (e.g., CO_2 emissions or water consumption). Clear and standardized labeling systems for eco-friendly products can facilitate the preference for sustainable options. The "EU Eco-Label" for textile and clothing products already exists.

Better washing and drying instructions: Washing and drying clothes hold a significant place in terms of environmental impact. Therefore, the industry can provide consumers with better information on reducing this impact. For example, care labels may include eco-friendly care instructions. Some companies also offer step-by-step repair and maintenance guides on their websites.

5 Sustainable Business Methods for Textile and Fashion

In the last forty years, the sustainability agenda has significantly evolved, prioritizing cooperation between governments and organizations within society by focusing on the implementation of strategies to address sustainability-related issues. Especially the Paris Agreement has been the first example where governments are directly held accountable for promoting such collaborations to battle climate change and environmental degradation.

Companies could play a critical role in advancing sustainability by adopting responsible practices aimed at reducing the environmental impact of their activities. These initiatives not only contribute positively to stakeholder relationships but can also create growth opportunities, reduce operational costs, and provide a competitive advantage.

The successful integration of sustainability strategies into business operations requires the transformation of existing business models. Sticking to traditional structures can hinder innovation and lead to strategic failure if leadership cannot adapt to sustainability requirements [31–34]. Fashion industry has been accused of taking limited precuations for sustainabilty- related issues considering the climate change discussion and over-consumption of natural resources. The industry is common in low-cost countries where ecological and safety regulations are not applied properly. For instance, one pair of pants and a t-shirt requires 19 L of water, demonstrating how much water is used in the production of cotton. The life cycle of clothing is regulated by the sector [31–34].

The environmental challenges associated with the fashion industry also encompass a significant waste management problem. The sector exerts control over the clothing lifecycle, where garments that are still relatively new are discarded—not due to physical deterioration, but because they are deemed unfashionable as a result of aggressive marketing strategies. This issue has prompted growing concern among industry stakeholders, leading to calls for more environmentally responsible practices. In particular, the industry is urged to reconsider its continued dependence on linear business models, which often overlook or externalize environmental costs. These traditional linear systems are characterized by a "take, make, dispose" pattern, which is increasingly viewed as unsustainable in the context of global environmental challenges [35–40].

The fashion sector is a huge and multifaceted industry which may be categorized into fast fashion and slow fashion segments. Scholars have emphasized the key distinctions between two models: fast fashion and slow fashion. Fast fashion is characterized by rapid production cycle, low-cost items, and high marketing strategies. On the other hand, slow fashion prioritizes craftmanship and durability over volume. The product development timelines may extend up to nine months. Nevertheless, even brands operating within the slow fashion paradigm face criticism for issues like overproduction and for destroying unsold inventory rather than donating it. Those concerns are becoming progressively significant in the European context [35–40].

Business models are fundamental structures that describe how organizations create, deliver, and capture value, thereby guiding the development of effective strategies. Companies use business models both as a strategic tool and to define customer segments and evaluate the value they offer in order to focus on value creation, profitability, customer demands, and market positioning. Each company's business model creation process may be unique, but during the design phase, five key elements are generally considered: people, strategy, structure, process, and reward. According to Osterwalder and Pigneur, the first element, people, refers to the alignment of the company's human resources and employee skills, mindset, and corporate culture with the strategic direction [41]. Strategy encompasses how the organization is positioned in terms of performance and competition, its main activities, and target customer segments. Structure determines the type of organizational framework that will support the business model. The process involves the design of the model, its implementation, and its adaptation in accordance with both internal and external factors (such as market dynamics, sectoral pressures, trends, and macroeconomic conditions). The reward, on the other hand, ensures that employees are motivated and that goals are achieved through human resources practices.

To create a reliable and effective business model, these five components must work in harmony. The critical point in the design process is how the company will create its competitive advantage and how it will sustain that advantage. This is directly related to how the organization's core strategy differentiates itself from its competitors. Additionally, business models function as strategic tools that guide communication with internal and external stakeholders. It explains value chain processes such as product sales, distribution, delivery, and customer service. In a study conducted by Osterwalder, Pigneur, and Tucci (2005), business models are defined by a system consisting of four fundamental cornerstones and nine building blocks: product, customer interface, infrastructure management, and financial structure. Each building block plays a critical role in value creation. For instance, the customer interface focuses on issues such as who the target customer base is, which communication and distribution channels will be used, and how customer relationships will be managed. Each of these blocks should be customized considering elements such as the company's financial structure, partner network, core competencies, and overall value configuration [41]. The difference between linear and circular business models draws attention to the life cycle perspective involved in circular systems. Linear models generally refer to a "cradle-to-grave" process—from raw material extraction to the end of the product's life—while the circular model adopts a "cradle-to-cradle" approach. In this system, materials are continuously reused, transformed, or recycled within the production and consumption cycle, minimizing waste, and maximizing resource efficiency.

The Sustainable Consumption and Production (SCP) cycle is one of the key tools for businesses to develop sustainability and support the circular economy. This tool, consisting of nine main principles, promotes a circular way of thinking that supports environmentally friendly consumption, such as resource management, sustainable design, production and research efficiency, eco-labelling and certification, sustainable marketing and lifestyle, and waste management [41, 42].

6 Sustainable Solutions of Some Textile and Apparel Brands

H&M Sustainable Strategies

Some textile brands have significantly increased their focus on sustainability-related initiatives.

For example, H&M group includes eight clearly-defined brands—H&M, COS, Weekday, Cheap Monday, Monki, H&M Home, & Other Stories and ARKET. Together, their brands offer customers a wealth of styles and trends in fashion, beauty, accessories, and homeware—as well as healthy, modern food in selected stores [43]. Their vision and strategy are built on three key ambitious; (1) **100% leading the change** (such as promote and scale innovation, drive transparency, Reward sustainable actions etc.), (2) **100% Circular & Renewable** (such as a circular approach to how products are created and utilized, usage of only recycled or other sustainably sourced materials, a climate positive value chain etc.), (3) **100% Fair & Equal** (such as fair jobs for everybody, stewards for diversity and inclusiveness etc.). To make sure that their sustainability work is done methodically, their global sustainability department, which has over 30 professionals, is in charge of establishing plans, targets, goals, policies, and follow-up procedures. They have about 150 employees who work only on sustainability in their 20 production markets. Through capacity building initiatives and activities, these colleagues collaborate closely with their business partners to evaluate their performance in relation to their sustainability commitment and assist them in improving. Based on the plan established by our global sustainability department, each function within each sales market then establishes its own quantifiable sustainability objectives and initiatives. They have a dedicated sustainability manager in each sales territory that oversees the execution of our plan throughout the market to involve all coworkers and clients.

The company currently supports and collaborates with around ten companies including:

Worn Again: are developing a ground-breaking chemical separation and recycling technology for blended fibre textiles.

Re:newcell: have developed a unique technology for recycling textiles made from cellulosic fibres (such as cotton, lyocell and viscose) into a pulp, which is utilized as input for creating fully circular viscose or lyocell.

Treetotextile: are developing a new sustainable process to make textile fibres from forest raw material in a sustainable way.

Thread: is a marketplace focusing on personalization. By combining tips from real stylists with machine-learning algorithms, Thread reduces the millions of style options available by only suggesting things that are perfect for you.

Sellpy: is a service helping people to easily sell their unwanted clothes, giving fashion a second life.

The company also supports the transparency strategy which means knowing how and where the products are manufactured and sharing the information wherever possible. As such, supply chain transparency creates greater accountability for both the company and the business partners. They introduced "**ARKET**", the newest brand in the H&M group, in 2017. The company, which was founded on the fundamental values of sustainability and transparency, enables users to browse all ARKET goods on its website by material type and production country. ARKET is leading the market with a fresh and creative approach to making supplier information customer facing, with openness as a fundamental component of its business strategy.

Customers who visit the ARKET website are instantly given the option to choose and purchase products based on the products' country of origin through a straightforward search function. As a result, each ARKET product's tier 1 supplier information is shown to the customer directly next to the product. The production unit and supplier name are among the details provided. Customers who purchase specific products can also read supplier stories, which provide them a detailed look at the facility that produces the product. This degree of openness not only raises consumer awareness but also gives suppliers credit for their contributions to the brand's merchandise. Through the sustainable impact collaboration, H#M collaborates with corporate partners. The business has been rewarding its ambitious and high-achieving associates with training opportunities and more orders. Every quarter, the management team of H&M UK & Ireland organizes a clothing collection competition for all UK & Ireland stores to inspire coworkers to actively participate in our sustainability strategy. In order to determine who can gather the most unwanted clothing through the in-store H&M garment collection program, store teams compete with one another. Next, a selection, announcement, and reward are made for the winning store. Customers are urged to donate unwanted clothing and textiles, regardless of brand and condition, to any H&M shop in the world so they can be recycled, re-worn, or reused through the company's in-store garment collection program. Customers receive a voucher in exchange that they can apply to their subsequent H&M purchases. Customers are also encouraged to return empty beauty containers for recycling, in addition to having their unwanted textiles collected and re-worn, reused, or recycled through the H&M group's collaboration. Customers receive 10% off their subsequent in-store purchases in exchange for either sustainable action [43].

Circular models reduce waste and maximize resources. Less than 1% of the materials utilized in the clothing industry are recycled into new items when using the current linear model [41, 42]. However, under a circular model, resources are utilized for providing possible beforehand being recycled into new materials and products, which reduces waste and adverse effects.

Making the transition from a linear to a circular model is not only environmentally beneficial but also highly cost-effective. 87% of the fibers used by the industry to make products wind up in landfills after consumers have stopped using them. Each year, this amounts to a missed opportunity worth over $100 billion. The business improves the efficiency and value of its operations and goods by designing for circularity, maximizing resource utilization by using recycled or other sustainably sourced materials, extending product lifespan, and establishing a value chain that is climate positive [41, 42].

Hugo Boss's Sustainable Strategies

Hugo Boss aims to provide circular products which are made with renewable/recycled materials. The products are designed for longevity to keep resources in a cycle for so long as possible. Hugo Boss's circular styles fulfil all three criteria including renewable or recycled raw material, utilizing recyclable products, utilizing products designed for longevity. Digital product creation includes the transformation of physical product development steps into digital processes; from product design and digital prototyping to interactive product presentations based on 3D simulation and animation. By using 3D simulations to create digital prototypes, the Hugo Boss company has already been able to significantly cut down physical samples during product development. Through data analytics, their goal is to harness their data across the entire value chain to extract actionable insights. By accurately predicting outcomes and identifying emerging patterns, they can enhance their procurement strategies, reduce unnecessary waste, and lower their carbon footprint.

By 2030, their objective is to procure all natural raw materials—such as cotton and wool—exclusively through regenerative agricultural practices or closed-loop recycling systems. Regenerative agriculture represents an environmentally restorative method of raw material production, which emphasizes the enhancement of soil health, the promotion of biodiversity, and the sequestration of carbon, thereby contributing positively to ecosystem sustainability. The fundamental aim is to preserve soil integrity, recognizing it as the essential foundation for sustaining life. In the context of closed-loop recycling, only recycled textile waste—sourced from pre- or post-consumer materials—is utilized, particularly in the production of fabrics and linings.

Their strategy is built on a strong environmental, social and governance basis (ESG). Their ESG -core includes environmental core, social core and governance core which includes dedicated action fields. Considering environmental core; Hugo Boss is planning to use recyclable or reusable packaging as of 2026. Also as of 2026, they aim to use only packaging plastic that consists of at least 50% recycled material. They aim for 100% of the electricity used to come from renewable sources as well as 20% reduction of energy consumption in relation to area (m^2) compared to 2019. When the social core is considered, they aim for at least 75% employee satisfaction results in our annual employee survey and at least 75% of vacancies for the top two management levels to be filled with internal candidates. They aim for a share of at least 40% women in the top management level (below the Managing Board) and at least 50% women in the second management level. As the governance core is considered; They only use mohair that comes from Responsible Mohair Standard (RMS) certified sources. As of their summer season 2025, they only use certified cashmere in products with more than 60% cashmere content [44].

The company also would like to minimize the polyester and polyamide fibre in all fabrics and linings in the products. They have been looking for a true alternative to polyester and polyamide fibers. They have innovated cellulosic yarn that could serve as an alternative to the conventional synthetic fibres like polyester or polyamide as part of a long-term, strategic partnership with HeiQ (LSE: HEIQ). The HeiQ AeoniQ (TM) yarn can offer an alternative to polyester and polyamide fibers. While polyester and polyamide fibers are made from non-renewable oil-based raw materials, the HeiQ AeoniQ yarn comprises renewable raw materials. This technology can help prevent water pollution from microplastic fibers and preserve agricultural land, when replacing oil-based materials like polyester. Collaborations like this are critical to increasingly eliminate synthetic fibers and diminish their impact on the planet. The company released the first HeiQ AeonieQ styles in the market which now includes polo shirts, a sneaker and outwear styles. the HeiQ AeoniQ yarn comprises renewable raw materials. This technology can help prevent water pollution from microplastic fibers and preserve agricultural land, when replacing oil-based materials like polyester. Collaborations like this are critical to increasingly eliminate synthetic fibers and diminish their impact on the environment [40]. HUGO BOSS has already released various HEIQ AEONIQ™ styles, including polo shirts, outerwear styles, a sneaker, and tailoring. The BOSS x HeiQ AeoniQ™ performance shirts which was unveiled at the BOSS Open 2023 was awarded with the ISPO Award 2023. The seamless high-performance polo shirt is produced from HeiQ AeoniQ™ fibers made from certified wood pulp. This is one of many possible renewable raw materials from which the yarn can be produced. More HeiQ AeoniQ™ styles were also in the spotlight at the FW23 Fashion Show [44, 45].

7 Challenges to Sustainability in Textile and Fashion

Fashion markets are oversaturated because of the extremely effective mass manufacturing system. The globe is awash with unsold clothing in addition to brand-new fashion items and stores [46]. Discounts in clothing stores appear to be a constant occurrence. Furthermore, not all new clothing ever makes it to the market; some, due to their subpar quality, end up in landfills right out of the factory. Additionally, because the market is oversaturated with options, certain clothing items are never sold to customers from the store. The quantity of these unsold clothes that wind up in landfills is actually unknown. These unsold clothes are thought to make up as much as 5–10% of the entire fashion industry. In order to preserve the name of their brands in this dangerous industry, fashion businesses would prefer to trash unsold clothing rather than mark it down and put it on the market in excess [46]. The following recommendations are provided by Sandy Black (2011) for a sustainable fashion designer: design for the full life cycle of clothing (including use and disposal) [47]; recycle, upcycle, repair, and remodel clothing; reuse waste materials; create multipurpose clothing; use ecological materials; use mono materials; use new technologies; create longer-lasting products; create multipurpose clothing; and design for pleasure.

Cradle-to cradle

Stahel first proposed the cradle-to-cradle approach in the 1970s [48]. McDonough and Braungart (2002) [49, 50] did it again at the start of the new millennium. In accordance with this notion, a product is made and constructed to allow for numerous life cycles of the material or product [50]. According to the cradle-to-cradle principle, a product will either continue in its technological or biological life cycles—that is, it will either be composted or recycled into new materials—after its usage phase. This implies that all materials, chemicals, dyes, and auxiliary materials must be biodegradable, meaning they can be recycled or composted. This is a difficult task since, for example, only a small number of textile colors or fiber compounds may be composted without causing harm to the ecosystem [50].

Redesign

At the start of the twenty-first century, it has been fashionable to repurpose existing materials into new fashion items. Reuse and redesign are sometimes referred to as eco-efficiency solutions, despite the fact that they don't deal with the underlying issue of rising output and consumption [50]. Reuse and redesign are simple to embrace because they don't require alterations to existing customer behavior. Additionally, redesign has emerged as a popular and fashionable fashion design strategy due to the massive volumes of textile and garment waste we produce [50]. It is helpful to keep in mind during the redesign process that not all textile materials are made for clothing, which could make them uncomfortable to use or even inappropriate to wear against the skin [51].

Recycling

Recycling allows expired products to come to life again. Recycling is the process of turning a product into new fibers or materials [50]. Recycling could include both upcycling and downcycling. On the one hand, downcycling is the recycling of a used product into a lower quality or less functional product, thereby reducing the value of the material. As a result, filling is the primary usage for downcycled material. On the other hand, upcycling is the creative reuse of waste or used materials, transforming them into a more valuable or functional product, thereby increasing the value of the material or gaining a new value. The recycling method requires mono materials, meaning that the entire garment—including the zipper, buttons, and threads—is constructed from a single material [50]. Because of this, recycling as a single piece and material is made simple. Another idea is that all components are simple to disassemble, and this is how goods must be created. However, this is difficult to realize in garments since disassembly would be too expensive and time-consuming [50]. Almost all clothing is composed of fiber mixtures that are difficult or impossible to recycle, and mono-materials are also uncommon in fashion goods. Polyester is the material of choice if we accept the current consumption patterns and the enormous amount of textile and apparel waste that goes along with them. High-quality material could be produced by melt-spinning polyester. This strategy ties us into the manufacturing of polyester and the related oil production, even though we see savings in energy, for example, the process still requires a certain quantity of virgin polyester material [50, 51].

Materials

Many different natural and chemical fibers are used in the production of textile and fashion products in the world, and these fibers find a life form in many different textile products. Despite the emergence of novel materials, cotton and polyester are the most commercially significant fibers utilized in the fashion industry, making up over 80% of all fibers, with polyester usage increasing annually [50, 51]. It is crucial to have a variety of textile fibers since diversity is a key value in sustainable growth. Therefore, from a diversity perspective, the existing dominance of polyester and cotton can be viewed as problematic. Hemp, which is thought to be an eco-friendlier substitute for cotton, has recently grown in popularity in marketplaces for sustainable fashion. Hemp requires little insecticide, grows quickly, and yields long fibers. It can even be grown in colder climates and enhances the structure of the soil [50–52]. Additionally, hemp procedures can make use of environmentally friendly enzymes. Flax is a fiber that is comparable. It is impossible to determine whether synthetic or natural fibers are better for the environment. Every fiber carries a unique environmental load. Acrylic, polyester, viscose, and cotton require a lot of energy throughout the production process and making them into fibers requires a lot of water. Certain natural fibers have a significant environmental impact during cultivation, such as the use of water and pesticides in the production of cotton, and the land space used to grow them may be better utilized for food production. In this sense, organic cotton is a superior option because it requires less chemicals (fertilizers and pesticides) to grow [50–52]. In contrast to cotton, which is mostly downcycled, synthetic fibers, which are byproducts of petroleum production, can be recycled more readily into high-quality materials like polyester [50–52]. However, premium cotton that has been recycled has just recently hit the market. For instance, bio-based plastics and newly developed fibers have promise for the fashion industry in the future. Corn (zein protein), peanut protein (ardil fiber) and soybean protein, for example, can be used to make these fibers. Protein fibers can also be manufactured from casein protein of milk [51]. The ideal scenario would be for all of these novel materials to be agricultural byproducts, biodegradable, and derived from renewable resources. Wood pulp and trees grown especially for this purpose are used to make ecological viscose, modal, and lyocell [51]. Lyocell is renewable and biodegradable, and its processing has been designed to have minimal ecological impact [52]. Furthermore, Switzerland is where this environmentally friendly bamboo fiber was created [53].

Because they require more frequent washing and create more perspiration during usage than clothing made of natural fibers, some synthetic fibers have a greater environmental effect. Beaches and marine ecosystems are being contaminated by microplastics from washing machine effluent [51]. When clothing consisting of synthetic fibers, such as polyester, acrylic, and polyamide, is washed, these microscopic particles become loose [51–55]. This issue is particularly prevalent in polyester sportswear that is frequently cleaned. Textile materials have been grouped by the organization Made By based on how they affect the environment [50]. According to Made By classification, Class A fibers have the least impact on the environment and include recycled cotton, recycled nylon, recycled polyester, organic hemp, and organic flax [56]. The qualities of the material and how well it fits the specific wearing habits must be carefully taken into account when selecting fabrics for clothing. Assessing the material's aging process is also crucial to preventing it from seeming outdated too quickly. According to studies, premium leather and wool are materials that are thought to age beautifully [55, 56]. In contrast, because of the tendency for pilling, materials such as acrylic appear aged after a relatively short period of use and this also occurs with soft fleece-type polyester, which is widely used.

The Use Phase

Washing, drying, and ironing textiles requires a lot of water and energy. It is known that just 7.5% of laundry is truly filthy. Most of them are cleansed more frequently due to behavioral or cultural factors [57]. The most substantial ecological impact is caused by frequently washed clothing. By maximizing the appropriate textile materials, fits, and hues for every purpose and use, both the consumer and the designer can reduce the quantity of washes. Compared to their anticipated moderate relative environmental impact during manufacture and disposal, clothing and workwear materials have a substantial relative environmental impact throughout use [51–57]. Consumer care can account for up to 75–80% of a cotton shirt's overall environmental effect, depending on the material and how often it requires to be washed [51–57]. In this instance, eco-friendly material selections don't really improve the environmental impact as measured by the entire Life Cycle Analysis (LCA).

Although reducing energy consumption is the aim of EU energy efficiency policies, new issues with clothes care are also emerging. The EU insists that every new washing machine has a temperature washing program set at 20 °C. Using cold water to wash clothes although water saves energy, harsher chemicals are required to remove all dirt from the fibres. Strong chemicals remain in the effluent, creating a new environmental issue. Additionally, using harsher chemicals during the launch process may exacerbate allergic reactions. Consequently, higher laundering temperatures rather than lower ones with stronger detergents could be preferable [58, 59]. The usage of chemicals in general is becoming a bigger issue in the fashion and textile sector. Numerous chemicals may be used in production processes, which are a danger to end customers as well as the employees of textile plants. Recently, there have been multiple product recalls involving clothing that included excessive amounts of dangerous substances. The REACH chemical legislation in the EU strictly limits chemical residues in clothing, textiles, and chemical waste from the textile sector. Consequently, there is a greater assurance that the clothing is safe for the final consumer if it is manufactured in an EU nation.

The lifespan of the garment is one of the most substantial determining factors of the ecological impact during the usage phase. Nowadays, clothes are much less expensive in contrast to household earnings from a few decades prior. Prices for apparel and textiles have decreased, and consumers now own an increasing amount of impulsively purchase low-quality, inexpensive clothing Textures. These types of inexpensive, low-quality clothing are simple to get rid of. One of the worthiest concerns for sustainable development is extending the life of clothing. According to research conducted in the UK, nearly half of people's clothing has been stored in closets without being utilized in the previous 12 months. This translates to an estimated 2.4 billion goods in the UK [46–52]. It's crucial to make investments in order to reduce consumption, excellent design, long-lasting style, and aesthetically pleasing aging materials in addition to good quality and durability. In this sense, services that try to prolong the lifespan of clothing have benefits in the sustainable context of development. One difficulty with the current approach is figuring out how to create products that are added with services that motivate customers to take more ecologically friendly actions. When focusing on ethical and sustainable use and consumption, the following factors should be considered: buying fewer clothes, investing in meaningful clothing, investing in long-lasting clothing, investing in high-quality, classical clothing, investing in eco-materials and eco-labels, prolonging the time that clothing is owned and being used more frequently, washing less, allowing clothing to rest and air between uses, maintaining clothing, also repairing, and using services to increase use and prolong the use time (e.g. repair, upgrading) [46–60].

New Digital Technologies

Compared to traditional industrial mass manufacturing, new digital technologies provide the potential for local production with less of an impact on the environment. Started manufacturing is made easier by new digital equipment, which don't require large investments locally. Digital technologies, such as 3D knitting and textile printing, allow for smaller production runs, the avoidance of surplus production, and even the ability to manufacture only in response to orders. Additionally, individual, and distinctive designs can be made through digital manufacturing. Additionally, clothing made nearby and locally can result in significant logistical savings.

The opportunity to do things differently is provided by digital textile printing. It offers the chance to produce locally, in small quantities or solely in response to orders, and with a customized and distinctive style.

The textile and fashion industry has experienced profound transformations over the centuries, evolving from manual craftsmanship to the integration of computer-aided design (CAD) and computer-aided manufacturing (CAM) systems [61]. This technological evolution has engendered a paradigm shift, enhancing precision, efficiency, and creativity within textile and apparel design and production processes.

The initial integration of CAD/CAM components into woven textile production was publicly introduced at the International Textile Machinery Association (ITMA) exhibition in 1983, held in Milan, Italy. Notably, AVL Looms UK, a subsidiary of AVL USA, showcased the first computer-interfaced handloom. During the same event, Bonas Machine Co. Ltd., a UK-based loom manufacturer, unveiled the first electronic Jacquard loom. Furthermore, pioneering CAD design systems for Jacquard textiles were presented by EAT/Designscope, Sophis Systems, and Scotweave. Among these, Sophis and EAT, in particular, established high benchmarks in the development of Jacquard weaving design software. EAT has arguably emerged as the global leader in Jacquard design systems, widely adopted by both industrial manufacturers and academic institutions. Its core strength lies in its ability to simulate woven textile structures in both two and three dimensions.

Additionally, NedGraphics, a Dutch company, introduced a carpet design system that utilized a large color monitor capable of displaying 256 colors. In the absence of digital color printers at the time, the NedGraphics system projected digital design images onto a white carpet displayed on the wall of their exhibition booth. Several early-stage (beta) versions of digital design systems tailored for textile printing, carpet design, Jacquard design, and knitting were also demonstrated at this event.

In the present context, CAD/CAM technologies enable designers to select yarns, sketch weave designs digitally, and transmit them rapidly to clients via printable or digital formats such as JPEG files [58]. Revisions can be executed swiftly in response to feedback. Technological improvements in monitor and printer calibration have facilitated the use of "virtual sampling" methods, although final production approvals are still typically required before full-scale manufacturing commences [62, 63]. While digital tools have significantly streamlined and accelerated the design process, modern weaving designers must be equipped with multidisciplinary competencies. These include a solid understanding of supply chain logistics, cost management, production techniques such as fabric analysis, pattern drafting, tension calibration, yarn length calculations, and proficiency in specialized CAD software packages [50–64].

One of the principal advantages of integrating CAD-CAM technologies in knitting lies in their capacity to significantly reduce production waste through precise digital manufacturing methodologies which creates sustainable knitting production.

By enabling numerical control over stitch distribution, CAD-CAM systems facilitate the generation of highly customized three-dimensional textile surfaces tailored to specific functional or aesthetic requirements. The intrinsic elasticity and structural flexibility of knitted fabrics further empower designers to experiment with innovative forms and functionalities that are often unattainable with conventional woven textiles, thereby expanding the creative and technical potential of textile design.

Recent research underscores the critical role of CAD-CAM systems in enhancing both the efficiency and effectiveness of the design process. These technologies streamline development workflows and bolster the competitiveness of textile products by supporting rapid prototyping, iterative design, and on-demand customization aligned with dynamic market trends [63]. Such adaptability is particularly vital in an industry continually shaped by shifting consumer preferences and accelerated fashion cycles. Furthermore, the integration of CAD-CAM systems enables the exploration of programmable textile materials capable of replying to environmental stimuli—such as changes in humidity—thus pushing the boundaries of what knitted structures can achieve in terms of functionality and responsiveness [63–68].

Traditional printing methods make extensive use of fabric, dyes, and chemicals. Before the print quality and colors are altered to look as they should, for instance, the industrial rotation printing technique requires several meters at the start of the print process (imperfect pre-printing can take even 15–20 m). In contrast, digital textile printing has advantages for the environment.

When opposed to rotation printing, digital textile printing has the following environmental advantages [69]:

- Up to 60% in energy savings; 80% in water savings; and 90% in ink savings
- a 90% reduction in color waste
- a 70% reduction in stock.

Digital printing reduces the need for dyes by accurately controlling the amount of dye applied to the fabric. Additionally, it is feasible to print only what is required or ordered, preventing overproduction. Additionally, a small quantity of textiles or even the patterns inside the garments might be printed. This preserves both fabric and dye [69].

8 Role of Technology and Artificial Intelligence on Textile and Fashion

Although the fashion industry is fundamentally defined as an industry based on the production of clothing and accessories, this definition falls short of fully explaining the distinctions between fashion and mere clothing. The transformation of a garment into a fashion item is not solely the result of design and production processes; it is also shaped by marketing and sales strategies. The fashion industry operates through a complex and multi-layered supply chain that extends from the procurement of raw materials to the fabrication of garments, and from distribution to the final consumer. The sustainability of this chain depends on the efficiency of production, design, management, and communication systems. Unlike other manufacturing sectors, the fashion industry is a dynamic structure that brings together numerous stakeholders including designers, shop owners, factory workers, tailors, artisans, media professionals, advertisers, wholesalers, and retailers. Furthermore, marketing, promotion, fashion shows, pattern making, textile designing, and consumers themselves are integral parts of this structure. In this context, fashion designers are required not only to possess the ability to design garments but also to adapt their designs to changing conditions.

Schwab [70], while describing how artificial intelligence surrounds us with innovations that are transforming our lives, highlights its impact across a wide spectrum—from autonomous vehicles and virtual assistants to translation software and drone technology. AI, enabled by the rapid growth in computational power and expanded data access, demonstrates its influence in numerous fields ranging from the development of new drugs to algorithms that predict individual interests. The scale of future transformations that AI may bring stays one of the most debated topics of our time. Luce [71], provides numerous examples of how AI is utilized within the fashion industry. These applications are examined under four main categories: shopping and product discovery, sales, design, and supply chain. A notable example in the context of supply chain innovation is the mobile application developed by the brand Badgley Mischka. Introduced during New York Fashion Week in 2018 [72], this application was available exclusively for iPhone and iPad users. Participants were able to give feedback on the runway collections through "like" or "love" options. The real-time data collected through the application was analysed to guide the brand's future design and production decisions. Such applications not only facilitate more accurate production planning but also help prevent excess inventory. The data collected allows for trend forecasting, cost optimization, and more attainable sustainability targets.

According to Elaine Stone [73], the fashion system can be analysed at four fundamental levels: the primary market, the secondary market, the retail level, and the auxiliary level. The primary market comprises the production of raw materials such as fibers, yarns, and fabrics. The secondary market encompasses design, collection development, garment production, and distribution processes. The third level involves retail and wholesale platforms such as boutiques, department stores, and e-commerce platforms. The auxiliary level includes media and consultancy tools that inform consumers about fashion products. The success of the fashion industry depends not on the performance of a single component but on the harmonious functioning of all these components. In particular, technological developments have profoundly transformed how fashion products reach consumers. Augmented reality-based marketing applications and the widespread availability of online shopping have facilitated consumer access and created new opportunities within the sector. As debates continue around whether artificial intelligence can truly produce art, the development of algorithms capable of creating visual artworks demonstrates the rapid progress in this domain. Applications such as AARON, Google Deep Dream, and GauGAN have emerged as significant examples in the field of visual arts. These advancements have also raised an important question in the context of fashion design: **"Can AI design?"** Artist and researcher Robbie Barrat [74] conducted a project to explore whether AI could generate a fashion show. Using data obtained from Balenciaga's runway presentations, Barrat trained a neural network model to learn the brand's design style and attempt to create a new collection. Although the AI was unable to fully replicate the depth and originality of human-led fashion design, it did demonstrate a remarkable ability to generate complex patterns and textures [74]. Additionally, the project culminated in a virtual fashion show using AI-generated garments and virtual models, showcasing the potential of AI-assisted creativity in fashion.

Another significant innovation in fashion design is the rise of wearable technology products. The concept of smart clothing not only fulfills the basic need for covering the body but also integrates functional features through conductive textiles that are sensitive to touch, pressure, or temperature. A notable example in this area is the "Levi's Commuter Trucker Jacket," developed in collaboration between Google and Levi's. First introduced in 2016 and released for sale in 2017, this seemingly ordinary denim jacket allows users to control their smartphone, music player, and navigation apps through touch-sensitive fabric panels embedded in the sleeves.

Thanks to conductive fabric technology and interactive design, this jacket represents a fusion of fashion and function. It is emblematic of how wearable technology is reshaping the design and purpose of everyday garments. These innovations underline the convergence of digital and textile technologies, indicating a future where garments are not only aesthetic but also responsive, data-driven, and multifunctional.

The ready-to-wear sector generates approximately 2 million tons of waste, 2.1 million tons of carbon emissions, and consumes 70 million tons of water annually contributing significantly to environmental degradation [75]. The industry's environmental damage includes excessive use of energy, raw materials, and water, in addition to fostering overconsumption, the rise of a throwaway culture, and the accumulation of discarded garments in landfills. Other critical issues include noise pollution caused by textile machinery and animal rights violations in the production of animal-based fibers, all of which indicate the disruption of ecological balance.

The fashion industry also presents challenges to social sustainability. The weakening of human values due to industrialization is among the primary concerns. In many Asian countries—such as China, Bangladesh, Vietnam, and India—low-cost labor is extensively used to increase profit margins in both fast fashion and luxury brands. However, this system fails to improve workers' living conditions. Many laborers, some starting as young as 12, earn less than $40 per month under harsh conditions. One of the most tragic examples of these injustices is the Rana Plaza disaster in Bangladesh, which occurred on April 24, 2013, claiming the lives of over 1,100 people. This tragedy revealed the urgent need for reform in the garment industry and raised awareness among consumers regarding social responsibility, ethical values, and human rights 2021. While the long-term effects of the COVID-19 pandemic on industries and consumer behavior have yet to be fully assessed, Dutch trend forecaster Li Edelkoort stated that the pandemic would not only slow the pace of daily life but also alter consumption habits. She coined the term **"quarantine of consumption"** to describe this phenomenon [76].

The pandemic-induced transformations in the fashion industry could be condensed under three major subjects:

1. An increased awareness of sustainability,
2. A reevaluation of the number of fashion collections and fashion weeks,
3. The acceleration of digitalization and the integration of artificial intelligence into fashion processes.

The growing role of digital tools in production and marketing has significantly reshaped the fashion industry, particularly during the COVID-19 pandemic. As homes turned into workspaces and outdoor activities were restricted, consumers began to reevaluate their consumption habits. Reduced spending deeply impacted the fashion sector, prompting a shift toward environmentally conscious behavior and slow fashion ideals. During this period, consumers not only focused on the appearance of garments but also began to care more about how those garments were created and the materials utilized in their manufacture.

The accumulation of unsold inventory led many brands to reconsider the frequency of their seasonal collections. Rather than launching 6–8 collections per year as is typical in fast fashion, many companies decided to produce fewer but more meaningful collections. For instance, Levi's launched campaigns encouraging consumers to buy fewer, higher-quality garments, while H&M highlighted collections made from recycled and sustainable materials. Digitalization has had a transformative effect on production, design, supply, and sales processes in the fashion industry. Technologies such as CAD-CAM systems have increased efficiency and speed in manufacturing, while zero-waste pattern-cutting techniques contribute to sustainability goals by minimizing material waste.

In the realm of retail, virtual shopping platforms offer personalized design options tailored to individual body measurements, creating a novel and interactive consumer experience. Moreover, the rapid growth of the digital fashion market—centered around 3D digital garments—demonstrates a new frontier where physical production is no longer required. With tools like CLO3D and Marvelous Designer, designers can create realistic, wearable digital clothes that are showcased and sold online. One of the pioneers in this space is **The Fabricant,** founded in the Netherlands in 2019 by Kerry Murphy. As the world's first digital fashion house, The Fabricant offers fully 3D-designed clothing for consumers. By eliminating the need for fashion shows, physical studios, and garment sampling, it presents a revolutionary model for both economic and environmental sustainability [63–77].

9 Conclusion

The textile and fashion industries are undergoing a profound transformation in response to growing environmental and social concerns. This chapter has examined a wide range of sustainable approaches, from raw material selection and production innovations to digitalization and the integration of artificial intelligence. It is evident that the industry's conventional linear models are increasingly being replaced by circular, transparent, and technology-driven systems that prioritize environmental stewardship and ethical responsibility.

As sustainability becomes a central focus, brands are embracing eco-friendly materials, adopting zero-waste production techniques, and implementing digital tools such as CAD-CAM systems and virtual design platforms. Furthermore, artificial intelligence is emerging not only as a tool for operational efficiency but also as a catalyst for creative innovation, particularly in design, retail, and supply chain management. These technological advancements, coupled with new business models that emphasize durability, recyclability, and consumer engagement, offer promising pathways toward a more responsible industry. Nevertheless, challenges remain—particularly regarding overproduction, excessive consumption, and the need for systemic change. Moving forward, a multi-stakeholder approach involving policymakers, manufacturers, designers, and consumers is essential to foster a truly sustainable fashion ecosystem. Only through collective efforts and continued innovation can the textile and fashion industries transition toward a future that respects ecological boundaries while preserving cultural values and social equity.

References

1. https://www.domestika.org/en/blog/11897-introduction-to-sustainable-fashion-what-is-it-and-why-is-it-important. Accessed on 20 March 2025
2. https://www.jeanologia.com/tr/sustainability/. Accessed on 10 March 2025
3. https://palomospain.com/?srsltid=AfmBOooIrr5iidHYfDP3LixankzD88FiTPYkBTzddM3vjhIBt_j5IPS. Accessed on 10 March 2025
4. Roy Choudhury AK (2014) Environmental impacts of the textile industry and its assessment through life cycle assessment. In: Roadmap to sustainable textiles and clothing: environmental and social aspects of textiles and clothing supply chain. Singapore, Springer Singapore, pp 1–39
5. Kozlowski A, Bardecki M, Searcy C (2012) Environmental impacts in the fashion industry: a life-cycle and stakeholder framework. J Corpe Citiz
6. European Parliament (2020) The impact of textile production and waste on the environment (infographics). European Parliament. https://www.europarl.europa.eu/topics/en/article/20201208STO93327/the-impact-of-textile-production-and-waste-on-the-environment-infographics
7. European Environment Agency (2023) Microplastics from textiles: Towards a circular economy for textiles in Europe. European Environment Agency. https://www.eea.europa.eu/en/analysis/publications/microplastics-from-textiles-towards-a-circular-economy-for-textiles-in-europe
8. https://www.europarl.europa.eu/RegData/etudes/BRIE/2019/633143/EPRS_BRI(2019)633143_EN.pdf. Accessed 20.04.2025
9. Global Fashion Agenda (2023) The GFA Monitor 2023. Global Fashion Agenda. Retrieved from https://globalfashionagenda.org/news-article/introducing-the-gfa-monitor-2023/
10. European Environment Agency (2022) Textiles and the environment: the role of design in Europe's circular economy. Publications Office of the European Union
11. Karthik T, Rathinamoorthy R, Ganesan P (2015) Sustainable luxury natural fibers—production, properties, and prospects. In: Gardetti M, Muthu S (eds) Handbook of sustainable luxury textiles and fashion. environmental footprints and eco-design of products and processes. Springer, Singapore. https://doi.org/10.1007/978-981-287-633-1_4

12. Farrer J, Finn AL (2015) Fashion and textiles design for wellbeing: value adding through practice-led transdisciplinary design research. In: Gwilt A (ed) Fashion design for living. Routledge, Oxon, pp 25–39
13. Farrer J, Walker M (2014) Establishing a live teaching model through funded research—BRIDGE Building Research and Innovation Deals for the Green Economy. In: Proceedings of the EDULEARN conference, Barcelona
14. Finn AL, Farrer J (2011) Sustainable fashion textiles design: value adding through technological enhancement. In: Fabricating the body: textiles and human health in historical perspective, University of Exeter, UK
15. Gersak J (2014) Physiological responses of human body as criterion for estimating the thermal characteristics of a clothing system. In: Proceedings of the 7th textile bioengineering and informatics symposium, Hong Kong
16. Johnson S (2013) Communicating sustainability: lessons from public health. The guardian. www.theguardian.com/sustainable-business/communicating-sustainability-behaviourchange publichealth?goback=.gde_4730524_member_225431839. Accessed 15 Apr 2024
17. Kozlowski R, Baraniecki P, Barriga-Bedoya J (2005) Bast fibers (flax, hemp, jute, ramie, kenaf, abaca). In: Blackburn RS (ed) Biodegradable and sustainable fibers. Woodhead Publishing Limited, Cambridge, pp 36–88
18. Langenhove LV (ed) (2007) Smart textiles for medicine and healthcare: materials, systems and applications. Woodhead Publishing Limited, Cambridge
19. Liu L, Yao J (2012) Properties of biocomposite fibers from cellulose nanowhiskers and cellulose matrix. J Fiber Bioeng Inform 5(2):207–215
20. Montenay F (2009) Transmission at the heart of luxury. In: Camilleri A et al (eds) The linen book, vol 29. CELC Masters of Linen, Paris
21. Mridula D, Singh KK, Barnwal P (2013) Development of omega-3 rich energy bar with flaxseed. J Food Sci Technol 50(5):950–957
22. Oomah DB (2001) Flaxseed as a functional food source. J Sci Food Agric 81:889–894
23. Orr J (1915) The international standard bible encyclopedia, vol 1. The Howard-Severance Company, Chicago, p 876
24. Paris MS (2014) Disillusioned hedonist shoppers. The economist online. http://www.economist.com/blogs/schumpeter/2014/02/luxury-goods-market. Accessed 27 Apr 2015
25. Piaget J (1972) The epistemology of interdisciplinary relationships. In: Berger G, Briggs A, Apostel GML (eds) Interdisciplinarity. Problems of teaching and research in universities. OECD, Paris, pp 127–139
26. Blackburn RS (ed) (2005) Biodegradable and sustainable fibers. Woodhead Publishing Limited, Cambridge
27. European Commission, Joint Research Centre (2014) Environmental improvement potential of textiles (IMPRO-textiles). Publications Office of the European Union. Retrieved from https://www.scribd.com/document/414328329/M1-RO-SJS
28. Šajn N (2019) Environmental impact of the textile and clothing industry. European Parliamentary Research Service. https://www.europarl.europa.eu/RegData/etudes/BRIE/2019/633143/EPRS_BRI(2019)633143_EN.pdf
29. https://saxcell.com. Accessed on 11.02.2024
30. Günaydın GK, Avinc O (2022) A sustainable alternative for the woven fabrics:"Traditional Buldan handwoven fabrics". In: Handloom sustainability and culture: entrepreneurship, culture and luxury. Singapore, Springer Nature Singapore, pp 87–117
31. Global Fashion Agenda (2024) The GFA Monitor 2024. Retrieved from https://globalfashionagenda.org/resource/the-gfa-monitor-2024-update/
32. Lozano R (2015) A holistic perspective on corporate sustainability drivers. Corp Soc Responsib Environ Manag 22(1):32–44. https://doi.org/10.1002/csr.1325
33. Nidumolu R, Prahalad CK, Rangaswami MR (2009) Why sustainability is now the key driver of innovation. Harv Bus Rev 87(9):56–64
34. Schaltegger S, Hansen EG, Lüdeke-Freund F (2016) Business models for sustainability: origins, present research, and future avenues. Organ Environ 29(1):3–10. https://doi.org/10.1177/1086026615599806

35. De Brito MP, Carbone V, Blanquart CM (2008) Towards a sustainable fashion retail supply chain in Europe. ELSEVIER 114:534–553 [CrossRef]
36. Nordic Fashion Association. Background. Available online: www.nordicfashionassociation.com/background (accessed 1 June 2016)
37. EcoWatch All Shades of Green. Available online: http://www.ecowatch.com/fast-fashion-is-the-second-dirtiest-industry-in-the-world-next-to-big--1882083445.html (accessed 30 Nov 2016).
38. Kerr J, Landry J (2017) Pulse of the Fashion Industry. In: Copenhagen Fashion Summit 2017; Global Fashion Agenda & The Boston Consulting Group: Copenhagen, Denmark, pp 2–134. Available online: https://www.copenhagenfashionsummit.com/wpcontent/uploads/2017/05/Pulse-of-the-Fashion-Industry_Executive-summary.pdf (accessed on 29 Jan 2017)
39. Blowfield M, Murray A (2008) Corporate responsibility, 3rd ed.; Oxford University Press: Oxford, UK, pp 1–413
40. Ellen MacArthur Foundation (2013) Towards The Circular Economy. Availableonline: https://www.ellenmacarthurfoundation.org/assets/downloads/publications/Ellen-MacArthur-Foundation-Towards-the-Circular-Economy-vol.1.pdf (accessed on 17 Nov 2017)
41. Osterwalder P, Tucci Osterwalder A, Pigneur Y, Tucci CL (2005) Clarifying business models: origins, present, and future of the concept. Commun Assoc Inf Syst 16:1–26
42. Akenji L, Bengtsson M, Briggs E, Chiu A, Daconto G, Fadeeva Z (2015) Sustainable consumption and production. In: A handbook for policymakers (Global Edition)
43. Source H&M Group Sustainability Report, Accessed on 17042025
44. https://group.hugoboss.com/en/sustainability/product/more-sustainable-materialsplanet
45. Hugo Boss, Sustainability Catalogue. https://group.hugoboss.com/fileadmin/media/pdf/sustainability/HUGO_BOSS_Sustainability_Booklet.pdf. Accessed on 15.05.2025
46. Niinimäki K (2013). Sustainable fashion: New approaches. Aalto University
47. Black S (2012) The sustainable fashion handbook. Thames & Hudson
48. Lovins LH, Lovins AB, Hawken P (2008) Natural capitalism: Creating the next industrial revolution. Back Bay Books
49. Lovins HL (2008) Climate: making sense and making money. In: Jackson T (ed) Sustainable Consumption. Earthscan, London, pp 37–50
50. McDonough W, Braungart M (2002) Cradle to cradle: Remaking the way we make things. North Point Press
51. Fletcher K (2008) Sustainable fashion and textiles: design journeys. Earthscan
52. Chen HL, Burns LD (2006) Environmental analysis of textile products. Cloth Text Res J 24(3):248–261. https://doi.org/10.1177/0887302X06293065
53. Steffen R, Wolf M, Lenz J (2013) Sustainable fibre processing: Innovations for eco-friendly textile production. Textile Outlook Int 165:32–39
54. Browne MA, Crump P, Niven SJ, Teuten E, Tonkin A, Galloway T, Thompson R (2011) Accumulation of microplastic on shorelines worldwide: sources and sinks. Environ Sci Technol 45(21):9175–9179. https://doi.org/10.1021/es201811s
55. Niinimäki K (2010) Eco-clothing, consumer identity and ideology. Sustain Dev 18(3):150–162. https://doi.org/10.1002/sd.455
56. Made-By (n.d.) Environmental benchmark for fibres. Retrieved from https://www.made-by.org
57. Cooper T (2005) Slower consumption: Reflections on product life spans and the "throwaway society." J Ind Ecol 9(1–2):51–67. https://doi.org/10.1162/1088198054084671
58. Lewis H, Gertsakis J (2001) Design + environment: a global guide to designing greener goods. Greenleaf Publishing
59. Vaatteiden kylmäpesu voi lisätä kemikaalien käyttöä (2013) YLE Uutiset. Retrieved from https://yle.fi/uutiset/
60. Belz FM, Peattie K (2011) Sustainability marketing: a global perspective (2nd ed.). John Wiley & Sons
61. Tao X, Zhang H, Shen Y (2016) Advances in functional textiles and design engineering. In: Tao X (ed) Handbook of smart textiles. Springer, pp 1–25. https://doi.org/10.1007/978-981-4451-45-1_1

62. Hao L, Ni Y (2022) Digital transformation in textile and fashion design: theory and practice. Textile Press
63. Sinha P (2020) CAD/CAM in textile and apparel design: a practical guide for professionals and students. Bloomsbury Publishing
64. Collier BJ, Collier JR (1990) Designing apparel through the flat pattern. Prentice Hall
65. Dwivedi R, Dwivedi M (2013) Application of CAD in textile and fashion design. Int J Sci Res Publ 3(12):1–4
66. Singh S, Arora M, Bahl R (2022) Role of computer-aided design (CAD) and computer-aided manufacturing (CAM) in fashion and textile industry. J Textile Eng Fash Technol 8(1):12–16. https://doi.org/10.15406/jteft.2022.08.00287
67. Indrie L, Bişboacă R, Cehan V (2019) CAD-CAM technology in textile industry and its impact on product development and innovation. Annals Univ Oradea, Fascicle of Textiles, Leatherwork 20(1):55–60
68. Sinke M (2023) 3D knitting and digital fabrication in textile design: Integrating CAD-CAM for sustainable and adaptive production. Textile Futures Press
69. Digital inkjet textile finishing (2013) International Dyer 198(2):35.
70. Schwab K (2017) The Fourth Industrial Revolution. Crown Publishing Group
71. Luce D (2018) Artificial intelligence for fashion: How AI is reshaping the fashion industry. Fashion Futures Press in Artificial Intelligence for Fashion, New York
72. Fashion Revolution (2020) Fashion transparency index 2020. Retrieved from https://www.fashionrevolution.org
73. Stone E, Farnan SA (2018) The dynamics of fashion. Bloomsbury Publishing USA
74. Barrat R (2018) Neural networks and fashion design: exploring AI-generated couture. Retrieved from https://robbiebarrat.com
75. McQuillan H (2020) Digital 3D design as a tool for augmenting zero-waste fashion design practice. Int J Fashion Design Technol Educ 13(1):89–100
76. Peacock SF (2012) IBM Watson: The future of Bollywood fashion is cognitive. IBM. https://admin02.prod.blogs.cis.ibm.net/blogs/think/2017/07/42019/
77. URL2 (2021) "The Fabricant" 3D example. https://www.thefabricant.com/. 14.06.2021

MIX
Papier aus verantwortungsvollen Quellen
Paper from responsible sources
FSC® C105338

If you have any concerns about our products,
you can contact us on
ProductSafety@springernature.com

In case Publisher is established outside the EU,
the EU authorized representative is:
Springer Nature Customer Service Center GmbH
Europaplatz 3, 69115 Heidelberg, Germany

Printed by Libri Plureos GmbH
in Hamburg, Germany